Geometry: An Investigative Approach

Phares G. O'Daffer
Stanley R. Clemens

Illinois State University
Normal, Illinois

Geometry: An Investigative Approach

Addison-Wesley Publishing Company
Menlo Park, California
Reading, Massachusetts • London • Amsterdam
Don Mills, Ontario • Sydney

This book is in the
ADDISON-WESLEY INNOVATIVE SERIES

ISBN 0-201-05420-5
ABCDEFGHIJ-HA-798765

Preface

Geometry is a subject rich in history and broad in scope which can be approached from a wide variety of points of view. During the past two decades those revising the mathematics curriculum in the schools of the United States have taken a point of view which has placed a strong emphasis on correct terminology and on the formal axiomatic, deductive structures of geometry. This emphasis has appeared even in courses for teachers at the elementary and junior high school level. This traditional, almost exclusive, emphasis on the deductive process has often led teachers to what we feel is a narrow view of geometry and has limited their ability to view geometry creatively and to enjoy geometric activity. Some teachers have even concluded that geometry is a sterile, uninteresting subject which should be given very little emphasis in the classroom.

We feel that deductive proof has been so strongly emphasized that the important role the inductive process plays in mathematics has been slighted, if not ignored. Mathematical relationships are often discovered inductively. Once they are discovered, it is the deductive process (proof) which is used for logical verification. Since traditional courses for teachers are short on interesting experiences to be approached inductively, it follows that most teachers enter the classroom inexperienced in a process which many feel should be central in their teaching, particularly at the elementary and junior high school level.

In contrast to many available geometry texts for teachers we have attempted to provide extensive discovery experiences and have not placed emphasis on the deductive process. We have also attempted to select aspects of geometry that are inherently interesting to teachers, and which spotlight the creative aspects of the subject. A common remark from teachers who have used this material is that most of the ideas were new to them and not a rehash of their previous experiences in high school geometry. Finally, we have chosen content that we feel is central to the topics currently being suggested for inclusion in the school curriculum.

Although our main thrust in this text is with geometric content, we feel that it is essential to keep in mind sound pedagogical techniques appropriate to the level at which this content is to be presented. Furthermore, we believe that teachers often tend to use teaching methods similar to the ones employed by their teachers. Consequently, we have attempted to develop this book so that its use will provide a good example of sound pedagogical technique. For example, one assumption which has guided us in our writing is that teachers are best prepared to direct discovery experiences for pupils if teachers themselves have had many similar experiences while in the student role. Students are encouraged to explore geometrical ideas using constructions, laboratory materials, and a variety of other investigative techniques. This emphasis on investigations—designed to set the stage for discovery of key geometrical relationships—is central in each of the chapters of the text and helps teachers gain confidence in their search for patterns in geometry. Students who have completed courses using a preliminary edition of this text often report that their attitude toward mathematics and their perception of geometry have improved as a result of the experiences they have encountered. In addition to the content development and the pedagogy suggested by the very nature of the investigations, pedagogical remarks and pedagogical activities are included throughout the book to provide additional practical ideas for teachers.

We feel that experiences in this text will be useful to a wide variety of persons. The material is especially appropriate for undergraduate students who are planning a teaching career at the elementary and junior high school level. We also feel that prospective secondary teachers can profit from the type of experience this text provides, especially those who might be teaching geometry courses for non-college-bound secondary school students or those who desire a course providing a multitude of motivational or enrichment ideas.

Also, the self-contained nature of many of the chapters makes the material useful for workshops for inservice teachers. Thus there are many settings in which this book can be used for both preservice and inservice education in geometry. Finally, we feel that the development and bibliography for several topics (tessellations, polyhedra, and transformations, etc.) are complete enough to make this book a valuable reference for teachers of mathematics at various levels.

This text lends itself to many different teaching situations and methods. The central role of the "investigation" facilitates a student-centered course, rather than the more traditional instructor-centered course. We have found that students react positively toward working on the investigations in groups of from two to four persons; they can test ideas and get immediate feedback from each other. As they become more familiar with the small group format, we find that it is a natural setting for the experience of sharing mathematical thoughts — an experience too often missed. As the students work on the investigations it is the job of the instructor

a) to ask penetrating questions and to give leading suggestions when necessary,
b) to communicate that learning sometimes involves making mistakes,
c) to emphasize the idea that a "good" question is often as valuable as a "good" answer, and
d) to encourage the student to look at a situation with an eye toward making a conjecture about it.

On the other extreme the text can be used in the large lecture situation. In this setting the investigation can be incorporated in the lecture or can be assigned as an individual student activity.

We have purposely included more material than can ordinarily be taught in a one-semester course. The chapters are related to each other, yet are independent enough so that it would be feasible to use chapters as individual units. The instructor using the material must take responsibility for choosing that which is to be utilized in the course. Chapter 1 is of an introductory nature and may be assigned for reading; it is intended to broaden the reader's conception of geometry. Some instructors may wish to place considerable emphasis on the measurement process and measurement found in Chapters 10 and 11 while others may view this material with less interest. Accordingly there exist several options for using this text. One approach is to follow the sequence of chapters as outlined in the table of contents and proceed through as much of the book as possible. An alternative is to begin the course with Chapters 10 and 11 and to follow with Chapters 2, 3, 4, and 5 in that order. Another sequence is to begin with Chapters 1 through 4, follow with Chapters 10 and 11, and then to complete the course by selecting material from among Chapters 5 through 8—all of which deal with transformations. Chapter 7 has been written so that it depends upon Chapter 5 but not upon Chapter 6. So Chapter 6 may be omitted without jeopardizing any later chapter. Some of the more difficult sections have been starred and may be omitted in a minimal course. Exercises which are more involved or open-ended have also been starred.

An option also exists for various degrees of pedagogical emphasis in a course using this text. On one end of the scale, an instructor might deemphasize the investigation, emphasize the content and the more difficult exercises, and omit all discussion of pedagogy. On the other hand if the instructor focuses primarily on the investigation and the pedagogy discussions, a course with far greater orientation toward methods of teaching school geometry would result. Our experience suggests that even in a more content-oriented course, it is important and motivational to allow the pedagogical comments to play some role. A few experiences with some of the suggested pedagogical activities appear to better equip the teachers to use new approaches in their own teaching. We have found that teachers can find satisfaction in developing new geometric experiences for children even when such experiences may not be included in elementary school texts.

May 1975   P. G. O'Daffer
Normal, Illinois   S. R. Clemens

Acknowledgments

We wish to express our appreciation to the elementary and secondary preservice and in-service teachers at Illinois State University who have used this manuscript in preliminary form and to Garth Runion who taught a preliminary version of the course. We gratefully acknowledge that they have provided us with ideas and valuable feedback.

Also we are indebted to Jane Frieden, Mary Beth Ulrich, Lois Ryan, and Marilyn Parmantie for their excellent work in typing the manuscript and preparing preliminary drawings on ditto. Also to Robert Ritt and Wilson Banks for their support and encouragement of the project. We also extend our gratitude to the authors of the many excellent articles and books on geometry which sparked our ideas, to the reviewers of our manuscript for their helpful reactions, and to the production staff at Addison-Wesley for their dedicated work in the preparation of this book.

Finally we wish to thank our wives Harriet and Jo for their unending patience and support during the writing of this manuscript.

Note to Persons Using this Book

I hear and I forget
I see and I remember
I do and I understand

This oft-quoted Chinese proverb indicates what we hope is the spirit of this text. The book has been designed to encourage you to become involved in exploring the ideas of geometry. These ideas relate to the world of nature, the world of art, and to the world of mathematical thought. Geometry can be exciting to study, and it is our hope that your experiences with this text will help you sense this excitement and develop confidence in extending your knowledge of the subject.

Involvement in investigation throughout the book may be a somewhat different approach to learning than that to which you are accustomed. At first it may seem strange to have a mathematics course which is more than someone explaining how to do some skill and then testing you on that skill. It may also initially seem unsettling to be asked to explore a situation to see "what you can learn about it" rather than be asked to arrive at a single answer to a problem.

Our experience has been that as people become involved in the investigations and work with laboratory materials such as a ruler, compass, mirrors, dot paper, geoboards, styrofoam solids, and so on, the ideas of geometry that were previously unclear and uninteresting begin to have real meaning and to arouse a curiosity which leads to further exploration. Also we have evidence that the investigations foster a new attitude toward what geometry is all about and a new confidence in a person's approach to a mathematical situation or problem.

We urge you to "let yourself go" and follow the dictates of your curiosity and interest in the investigation situation. Also, when it is possible to work in groups, the involvement with others in a setting where you can test your ideas on them and give and get feedback can be a very valuable learning experience.

It has been said that a "good" question in mathematics is sometimes as valuable as a "good" answer. Real involvement means reaching the stage where you begin to ask questions and even suggest conjectures. This type of involvement often takes you beyond the basic assignment and into some articles, books, or experiences related to your interest. It is our hope that you can find the satisfaction of this type of involvement from this text.

We encourage you to enter the experience of the text with an open mind and with the curiosity of a small child. If your experience translates into an approach to teaching that places even a small premium on helping your students develop a spirit of inquiry, a confidence in approaching new situations, and a clear knowledge and deep appreciation of geometry, then our goals for writing the text will be in the process of realization.

P. G. O'Daffer
S. R. Clemens

Contents

**Chapter 3
Patterns of Polygons
in the Plane**

**Chapter 4
Polyhedra and
Tessellations of Space**

**Chapter 5
Motions in the
Physical World**

* An asterisk beside a section title indicates that the section is less basic
to an understanding of the material that follows, or is more difficult than usual
and may be omitted in a minimal course.

1

A Panoramic View of Geometry

INTRODUCTION

In the fourth century B.C. the Greek philosopher and teacher Plato (429–348 B.C.) carved the inscription "Let no one ignorant of geometry enter my doors" above the entrance to his academy. From that time to the present the question "What is geometry?" has been appropriate for both students and teachers. The material in subsequent sections of this chapter provides information which sheds light on an answer to this question. Before reading further, however, persons interested in teaching geometry might find it valuable to analyze their own ideas and views about the nature of geometry and to explore the ideas of others. The following "Investigation" suggests group activities which might be used.

In the next few sections we shall look at geometry in the physical world, geometry as a mathematical system, and the

---

Investigation 1.1

Join a small group and together engage in one or more of the following activities. Don't feel you must "know all the answers" to participate, but instead think of the group situation as a safe place to share untested ideas and to ask questions.

Activity 1 Write not more than five key words which best suggest your ideas of a general answer to the question "What is geometry?"

Write as many words as you can that suggest basic concepts that are studied in geometry.

Activity 2 Draw a picture which expresses symbolically your group's answer to the question "What is geometry?" The quality of the art isn't important, but the components of the picture should convey the intended message.

---

aesthetic and recreational aspects of geometry. It is hoped that after reading these sections, the reader will have made progress in formulating an answer to the question "What is geometry?"

GEOMETRY IN THE PHYSICAL WORLD

As we consider the nature of geometry, we cannot help but imagine when, where, why, and how man first became involved with geometric ideas. When we become sensitive to objects such as seashells, spider webs, honeycombs, snowflakes, sunflowers, and stars, we begin to see that geometric form and structure permeates the universe, and that man was immersed in a geometrical setting from the beginning. It remained for him to notice it, to appreciate it, to abstract ideas from it, and to use it.

Activity 3 Describe at least one specific example for each of the following aspects of geometry. If your group thinks no examples exist, mark "N." If you think examples exist, but you can't think of any, mark "C."

   a) Geometry in nature
   b) Geometry in art
   c) A modern-day prac-
      tical use of geometry
   d) Geometry as a theory
   e) An historic practical
      use of geometry
   f) A recreational use of
      geometry
   g) A scientific application
      of geometry

It is important for teachers at all levels to realize the crucial effect their attitude toward a subject has on their students. An adequate perception of the breadth and scope of geometry can positively affect your attitude toward geometry and hence affect the attitude of your students.

Nautilus shell

Honeycomb

Fern leaf

Snowflake

Pine cone

Sunflower

At all age levels, a person's environment is a rich source of geometrical ideas. Preschool children should be encouraged to notice balls, blocks, cans, cones, dials, and a whole universe of other shapes. Older students might be challenged to investigate the geometry of sunflowers, snail shells, starfish, bridges, architecture, Wankel engines, and other intriguing situations. The Nuffield Project book *Environmental Geometry*, John Wiley & Sons, N. Y., 1969, has some suggestions for teachers.

A study of the Golden Ratio provides an interesting setting for enrichment activities for older students. Ideas involved are: ratio, similarity, sequences, constructions, and other concepts of algebra and geometry. A book by Garth Runion entitled *The Golden Section and Related Curiosa* (Scott Foresman, 1972) provides some interesting extensions of this idea and contains a useful bibliography.

It is difficult to pinpoint the persons, the times, or the places where the intricate geometry of nature was first noticed by humans. For example, it has been observed that the ratio of a to b in each of these pictures is close to 1.618. This ratio, called the Golden Ratio, is the ratio of the length to the width of what is said to be one of the most aesthetically pleasing rectangular shapes. This rectangle, called the Golden Rectangle, appears in nature and is used by humans in both art and architecture. The Golden Ratio can be noticed in the way trees grow, in the proportions of both human and animal bodies, and in the frequency of rabbit births. Whoever first discovered these intriguing manifestations of geometry in nature must have been very excited about the discovery.

Snail

Parthenon

$$\phi = \text{ratio } a:b \approx 1.618$$

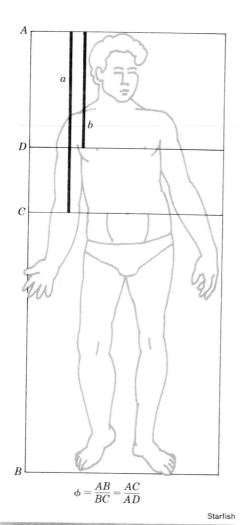

$$\phi = \frac{AB}{BC} = \frac{AC}{AD}$$

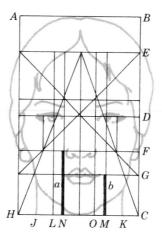

$$\phi = \frac{BC}{AB} = \frac{BD}{ED} = \frac{DC}{FC} = \frac{ED}{DF} = \frac{DG}{DF}$$

$$\phi = \frac{FC}{GC} = \frac{HC}{JK} = \frac{JK}{LM} = \frac{LM}{NO}$$

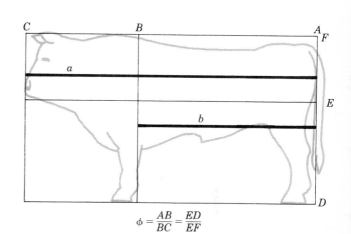

$$\phi = \frac{AB}{BC} = \frac{ED}{EF}$$

Starfish

Sunflower

Number of Type A spirals = a
Number of Type B spirals = b

The geometry of the physical world can be touched and seen. It is all around us. The child experiences it as he observes, handles, and alters; the architect experiences it when he contemplates and designs a building; the carpenter experiences it when he designs and constructs a cabinet; and the botanist experiences it as he examines and studies the shapes and patterns found in plant life. In a very real sense, one *experiences* physical geometry, and this experience is extremely important.

Early man's experience with geometric forms in nature was translated into useful solutions to his everyday problems. He appears to have reduced nature's shapes to simpler forms, which he used in various ways. For example, early man used geometric forms derived from nature to invent a method of pattern making which he used to decorate his utensils, and to make weapons and useful tools.

Arrowheads, baskets, and pottery exhibit geometric forms.

Pyramids at Giza

Later, as his understanding of physical geometry grew, he developed even more sophisticated uses of this knowledge. For example, since the Nile River annually overflowed its banks, the Egyptians used empirical "rule of thumb" geometric techniques to re-mark the land boundaries that had washed away. The word "geometry," which means "earth measure," originated from this setting.

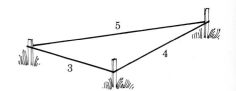

Perhaps one of the important early discoveries was that a triangle with side lengths 3, 4, and 5 is a right triangle. Thus, knots in a rope spaced 3, 4, and 5 units apart provided a useful method of determining a square corner for a field or for the base of a pyramid.

Historically, the experience with physical geometry was of great value. It served to provide practical methods, usually involving measurement, for solving the problems of the day.

But perhaps even more importantly, it provided the foundation for the more logical approach to geometry developed by the Greeks which was ultimately refined to become useful in scientific applications. In a similar way, experience with physical objects plays an important role in the child's learning of geometry, for this experience provides the necessary foundation for later abstraction and generalization of geometrical ideas.

Clearly, the geometry of actual experience—physical geometry—cannot be overlooked when answering the question "What is geometry?" for it has played a major role in both the history of geometry and in the curriculum for the modern elementary and junior high school student.

GEOMETRY AS A MATHEMATICAL SYSTEM

Geometry Organized Logically

It is appropriate to ask young children why a particular geometric relationship is true. While their response will likely involve a reference to measurement or other experimentation with physical materials, these "proofs," based on the physical world, provide a valuable background for a later understanding of proofs which rely on postulates and logic.

In physical geometry, points, lines, and geometric figures are considered in much the same way as we consider physical objects. We often rely on the appearance of figures or on approximate measurements in making decisions about geometric properties and relationships. The situations in Fig. 1.1 (a through f) suggest that appearance is sometimes deceiving. Perhaps it was situations like these, as well as a new emphasis on the value of reason, that caused Greek philosophers to develop a geometry that relied less upon visual information and more upon logical reasoning.

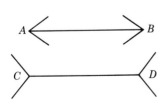

(a) Is it further from *A* to *B* or *C* to *D*?

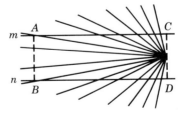

(b) Would line *m* meet line *n* if they were extended?

Figure 1.1

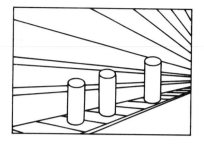

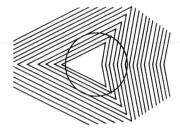

(c) Which cylinder is tallest?

(d) Is the figure in the center a perfect circle?

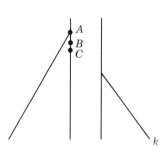

(e) Will line k, when extended, meet point A, point B, point C, or none of these points?

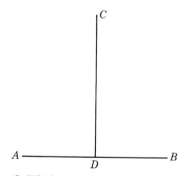

(f) Which segment is longer, AB or CD?

Regardless of the motivation, a new view of geometry emerged in the first half of the sixth century B.C. Thales of Miletus (640–550 B.C.) was one of the first Greeks to insist that geometric fact be established not by trial and error observation and experimentation, but by logical reasoning. His efforts set the stage for the monumental work of Euclid (300 B.C.) called *The Elements of Geometry*. Thales, Euclid, and other Greeks helped geometry evolve from the purely physical level to the more abstract logical level. Euclid's work has so influenced the teaching of geometry that, until quite recently, most secondary school students have thought of "geometry" and "proof" as essentially synonymous.

The spirit of this "Greek" view of geometry is illustrated by the story of Euclid's student who, when he learned the first theorem, asked Euclid, "But what advantage shall I get by learning these things?" Euclid called his assistant and said, "Give him threepence, since he must need to make a profit out of what he learns."

The development of geometry by Euclid, which seemed to disregard the practical, was essentially a logically derived chain of theorems based on the following initial postulates and common notions.

Postulates

1. A straight line can be drawn from any point to any point.
2. A finite straight line can be produced continuously in a straight line.
3. A circle may be described with any center and distance.
4. All right angles are equal to one another.
5. If a straight line falling on two straight lines makes the interior angles on the same side together less than two right angles, the two straight lines, if produced indefinitely, meet on that side on which the angles are together less than two right angles.

Common Notions

1. Things which are equal to the same thing are also equal to one another.
2. If equals be added to equals, the wholes are equal.
3. If equals be subtracted from equals, the remainders are equal.
4. Things which coincide with one another are equal to one another.
5. The whole is greater than the part.

GEOMETRY AS A FORMAL AXIOMATIC STRUCTURE

For two thousand years after Euclid listed the five postulates and five common notions which formed the basis for his logical organization of geometry, mathematicians were concerned about the need for the fifth postulate. The version of this postulate usually encountered is as follows:

Through a given point not on a given line there can be drawn only one line parallel to the given line.

Since Euclid's original statement of this postulate (see list of postulates above) is so much more complex than the first four, and since its converse could be proved, many mathematicians felt it could be proved using the other postulates.

The story of the futile attempts to find a proof, and the discoveries resulting from these attempts, presents one of the more exciting accounts in the history of mathematics. In the nineteenth century Gauss, Bolyai, and Lobachevsky discovered independently that new, logically valid, non-Euclidean geometries resulted when Euclid's fifth postulate was *replaced* by a contradictory postulate. From two thousand years of belief that there could be only one system of geometry to describe space evolved the conclusion that different sets of postulates, chosen to be consistent, could be used, and new legitimate geometries could be developed.

This discovery of non-Euclidean geometries, often classified as one of the most significant events to happen in mathematics, marked the beginnings of an interpretation of geometry as a formal axiomatic (or postulational) structure,* rather than a formal study of the physical world, and contributed greatly to an expanded view of the nature and use of geometry. In fact, Einstein in his 1921 lecture "Geometry and Experience" asserted, "To this interpretation of geometry I at-

Providing a good intuitive background in geometry at the elementary and junior high level makes it possible for secondary-level students to explore new topics. Chapter 8 of *A New Look at Geometry* (Signet Science Library, 1966) by Irving Adler provides ideas for special projects or units for the secondary-level student on non-Euclidean geometry.

---

* As a formal axiomatic structure, a system of geometry can roughly be described as consisting of the following:
  a) a set of elements and a collection of undefined terms;
  b) a set of statements called axioms (or postulates) which establish relations between the undefined terms, and which are accepted without proof;
  c) a set of definitions which utilize undefined terms or previously defined terms to specify additional concepts;
  d) a system of logic which provides a means for determining the truth or falsity of new statements within the system; and
  e) a chain of statements called theorems, each of which is logically deduced from axioms, definitions, or previously proven theorems.

tach great importance, for should I have not been acquainted with it, I would never have been able to develop the theory of relativity." Not only has the axiomatic view of geometry provided many useful models for scientific applications, but it has also had a developmental influence on many other areas of mathematics as well.

AESTHETIC AND RECREATIONAL ASPECTS OF GEOMETRY

Geometry as an Art

Throughout the centuries philosophers, mathematicians, and some artists have recognized and emphasized the artistic aspects of mathematics. G. H. Hardy (1877–1947), a well-known mathematician who taught many years at Cambridge University, said:

> A mathematician, like a painter or a poet, is a maker of patterns. If his patterns are more permanent than theirs, it is because they are made with ideas.
>
> The mathematician's patterns, like the painter's or the poet's, must be beautiful; the ideas, like the colors or the words, must fit together in a harmonious way. Beauty is the first test; there is no permanent place in the world for ugly mathematics.

When the Hindu mathematician Bhaskara (A.D. 1114–1185) drew the picture in Fig. 1.2 and offered no further explanation than the word "behold," he was not suggesting that the lines in his diagram were particularly appealing to the eye, but rather that this particular arrangement suggested a "beautiful" proof of the Pythagorean Theorem*—one that probably gave him a feeling of delight and satisfaction.

BEHOLD!

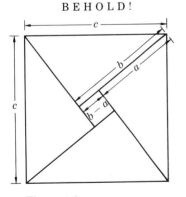

Figure 1.2

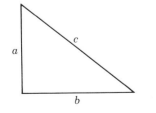

---

* From consideration of the area of the large square (c^2) in the figure above as formed by the area of the 4 triangles ($4 \cdot \frac{1}{2}ab$) and the area of the center square ($(b-a)^2$), we get the equation $2ab + (b-a)^2 = c^2$, which easily simplifies to $a^2 + b^2 = c^2$.

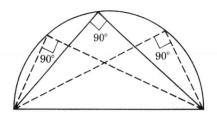

From the time that Thales was believed to have sacrificed a bull in joy over the discovery and proof that any angle inscribed in a semicircle is a right angle, mathematicians have found an inspiring beauty in the structure of mathematics.

As mentioned earlier, the discovery of non-Euclidean geometry and the resulting axiomatic approach to geometry brought this structure into clear focus. It is interesting to note that Johann Bolyai, one of the "codiscoverers" of non-Euclidean geometry, wrote a letter at that time to his father in which he said, "Out of nothing I have created a strange new universe." Perhaps it was in reference to the purity of geometry as an axiomatic structure that prompted the philosopher Bertrand Russell (1872–1969) to say:

> Mathematics possesses not only truth, but supreme beauty — a beauty cold and austere, like that of sculpture, without any appeal to our weaker nature . . . sublimely pure, and capable of stern perfection such as only the greatest art can show.

If man differs from other animals in being able to create and admire beautiful objects, then the mathematician joins the ranks of artistic man in that he is able to admire and create beautiful ideas. This beauty of geometrical ideas and the opportunity to participate, at an appropriate level, in the creation or discovery of "new" relationships is what has interested many children and adults in the study of geometry.

Geometry in Art

Another aspect of geometry as art is the actual explicit use of geometric principles or figures in the creation of designs, paintings, sculptures, or other art objects. Perhaps pictures, as presented on the following pages, better than words, can be used to illustrate this.

Perhaps man-made geometrical artistic creations cannot equal those which are found in nature, but when asking "What is geometry?" it seems unthinkable that one would omit the aesthetic aspect.

Students often find it interesting and valuable to create drawings and artistic expressions of geometric ideas. Curve stitching or ruler and compass designs provide useful activities in this regard. See the books *Line Design* and *String Sculpture*, published by Creative Publications.

Piet Mondrian, *Broadway Boogie Woogie*. 1942–43. Oil on canvas, 50" × 50". Collection, Museum of Modern Art, New York

Note the effective use of aesthetically appealing rectangles.

St. Jerome, DaVinci

Note the golden rectangular outline into which the main figure will fit.

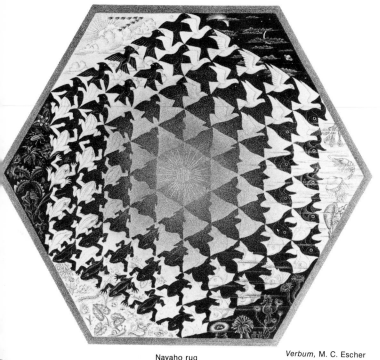

Navaho rug

Verbum, M. C. Escher

Modern sculpture

African sculpture

Quilt

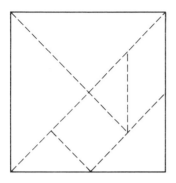

Geometry for Recreation

The recreational potential in geometry has been known and pleasurably exploited for centuries. Simple dissection puzzles in the plane and in space, for example, have long been a source of amusement. A few of these are given here.

1. Trace this square and cut it into seven pieces, as shown. You have the famous Chinese Tangram puzzle which is recorded to be at least 4000 years old.

   Can you place all seven pieces together to form the following shapes?

a triangle	a trapezoid
a rectangle	a pentagon
a parallelogram	a hexagon

2. An interesting book on tangrams* contains puzzles which challenge the reader to make all sorts of shapes with tangram pieces.

   Can you place all seven tangram pieces together to form the following shapes?

this "portrait"	a letter of the alphabet
a double arrow	an interesting figure
a digit	

3. Trace each of the following figures (a through g) and cut them into pieces as shown. Form single squares using all the pieces in each figure.

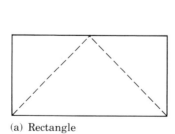

(a) Rectangle

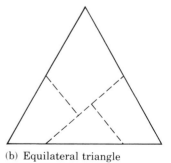

(b) Equilateral triangle

---

* Ronald C. Read, *Tangrams-330 puzzles*, New York: Dover, 1965.

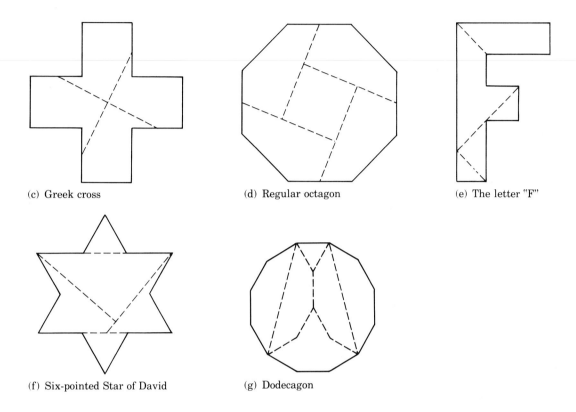

(c) Greek cross   (d) Regular octagon   (e) The letter "F"

(f) Six-pointed Star of David   (g) Dodecagon

4. Can you figure out how to make a model (from clay, wood, styrofoam, paper, etc.) of a space figure that is solid (not deformable) but which can be used to plug up completely each of these holes (a through c); and which, with a light tap, can be pushed through each?

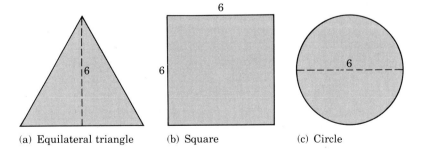

(a) Equilateral triangle   (b) Square   (c) Circle

Recreational geometry can be motivational. But teachers should also be aware that important concepts can sometimes be developed through recreational experiences. For example, the "proof" in Fig. 1.3, regarded as a puzzle, relates directly to area and the Pythagorean theorem.

Other recreations have involved searching for convincing proofs of important theorems. Henry Perigal, a London stock-broker and amateur astronomer, discovered the paper and scissors "proof" of the Pythagorean Theorem (Euclid's 47th theorem) described in Fig. 1.3.

Figure 1.3 Trace and cut out the triangle and the two squares on the legs. Draw two lines through the center of the larger square, intersecting at right angles with one of the lines perpendicular to the hypotenuse of the right triangle. Cut out the four pieces of the larger square and put them with the smaller square to form a large square on the hypotenuse.

In the spiral approach to instruction, important ideas are introduced early in simple form and are reintroduced and broadened at successive levels in the child's development. The puzzle pictured below can give young children an opportunity to experience the beginning aspects of the Pythagorean Theorem.

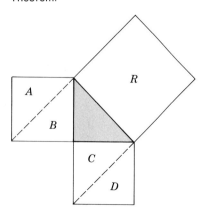

The solution to this "puzzle" shows that the area of the squares on the legs of a right triangle is equal to the area of the square on the hypotenuse, for example, $a^2 + b^2 = c^2$.

In a more formal way, many nonmathematicians have found a hobby in searching for proofs of the Pythagorean Theorem. James Garfield, the twentieth President of the United States, devised an ingenious proof for the theorem, which appears along with others suggested by both mathema-

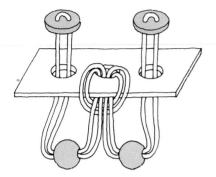

Figure 1.4 This puzzle is made from string, cardboard, two beads, and two buttons. The object is to undo the loop and get the two heads together without cutting or untying the string, tearing the cardboard, or taking off the buttons.

In this puzzle each person ties a piece of string to each wrist in such a way that the strings are looped as shown. The object is to separate from the other person without cutting the string, untying the knots, or taking the string off the wrists.

ticians and nonmathematicians in the book, *The Pythagorean Proposition*. Other puzzles, such as the topological puzzles illustrated in Fig. 1.4 and those appearing in subsequent chapters of this book suggest the variety of recreations available.

The recreational value of the problems and the many games and magic tricks based on geometric ideas is well-known.

These illustrations are only a few examples of the areas in which geometry has provided recreation for many. The literally hundreds of mathematical recreation books which have been published in the past 50 years, with titles such as *Play Mathematics, 536 Puzzles and Curious Problems,* and *Mathemagic,* attest to the popularity of mathematics in general, and geometry in particular as a recreational activity. This recreation aspect of geometry is not only "just plain fun," but it can also be used as motivational material for a serious study of geometry.

Many teachers are surprised to find that geometry can be interesting and fun for students of all ages. Beyond this, there are also many other practical reasons for teaching geometry at every level. Pages 28 through 50 and other sections of the Thirty-sixth Yearbook of the National Council of Teachers of Mathematics, entitled *Geometry in the Mathematics Curriculum,* suggest other reasons for teaching geometry.

SUMMARY

We began this chapter by asking, "What is geometry?" Although it is an oversimplification, we like the answer, "Geometry is the study of space and spatial relationships."

Some choose to emphasize in this study those relationships in the physical world which can be discovered through handling and observing physical objects and through the application of inductive reasoning. Others prefer to create an abstract mathematical model (system) and to study these relationships through the logic of the abstract system. This approach employs deductive reasoning and is the more traditional approach to geometry in the schools. Still others are content to study and experience these relationships in a much less structured way — through the aesthetic appeal of art objects.

Regardless of the approach chosen, a prospective teacher of geometry who studies the subject begins to realize that it has many dimensions. He realizes that a great deal of geometry emanates from certain physical situations and experiences. He realizes that geometry can be logically challenging, that it involves patterns and can be beautiful, that it is a fertile vehicle for creativity, and that it can be fun. It becomes clear that to really understand geometry, one must become deeply involved in these various aspects of the subject. Any one of the above aspects could easily be the central goal of an entire course. Ptolemy (?–A.D. 168) asked Euclid if there was an easier way than studying Euclid's "Elements" — a "king's shortcut" — to an understanding of geometry. Euclid replied, "There is no royal road to geometry."

In the chapters that follow, we have chosen not a "royal road" to geometry, but a particular path designed to provide the teacher of geometry with important experiences and valuable insights into the subject that he may not have encountered previously.

Along this path the reader will explore ideas such as the Golden Ratio, star polygons, tessellations of the plane and of space, motion geometry, topology, and many other aspects of geometry. It is hoped that through this exploration each person will be motivated to provide his own answer to the basic question of this chapter.

EXERCISE SET 1.1

1. Describe an aspect of geometry in nature that has not been discussed in this chapter.

2. Is a 3 × 5 card a Golden Rectangle? Explain.

3. The ratio of a person's height to the height of the person's navel is 172 to 106. By how much does this ratio differ from the Golden Ratio?

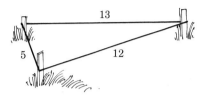

4. Suppose an unconventional Egyptian used a knotted rope, as shown here, to establish a square corner for building a pyramid. Would it work? Explain.

5. Euclid's geometry utilized postulates, definitions, and theorems. State a postulate, a definition, and a theorem which you remember from your previous study of geometry.

6. The idea of a formal axiomatic structure was inherent in the development of non-Euclidean geometry. If the "plane" in a certain non-Euclidean geometry is the surface of a sphere and a "line" is always a great circle of a sphere, how many lines will there be through point *P* not intersecting line *l*?

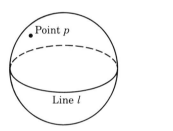

7. Create a "work of art" using geometric ideas. It need not be elaborate.

8. Complete the puzzles on pages 18 and 19.

9. Try the puzzles on page 21.

10. Invent a geometric puzzle.

BIBLIOGRAPHIC REFERENCES*

[4, 1–16, 195–260], [9], [15, 88–103], [26, 160–169], [34, 1–14], [40, 1–5, 152–162], [42, 89–102], [44], [53], [89], [98, 108–116]

---

* References to the bibliography at the back of the book are given at the end of each chapter. Page numbers are given when applicable.

PEDAGOGICAL ACTIVITIES FOR THE TEACHER

1. Create a project assignment for students at a level of your choice. One goal of the project should be to help the students become more aware of geometry in their environment. Write up a project description which you could mimeograph for the students or which you would use to describe the project to the students.

2. Devise a simple diagnostic interview which can be used to find out what a student at a given level of your choice knows about the ideas of geometry. You may wish to use geometric models, pictures, or devices and verbal instructions as part of the interview. Interview a student or group of students and summarize your findings.

3. Make a set of photographic slides depicting geometry in the environment. Possible settings for pictures are architecture, sidewalk patterns, nature, window designs, quilts, and so on. You may wish to tape music and synchronize this music with your slides. Present your slide-music program to a group of students. What was their reaction?

4. Geometric recreations can be valuable for motivating students. Refer to books which include geometric recreations and start a card file containing at least 5 such recreations which you feel confident would be useful to motivate students at a level of your choice.

2

Points, Lines, and Polygons

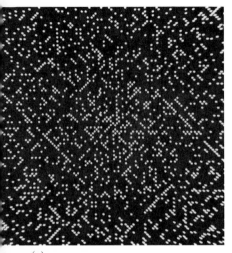

(a)

(b)

INTRODUCTION

Man has used his creativity either to discover in the physical world or to create for himself many geometrical patterns and designs. The reader likely has seen many interesting point, line, and polygon patterns like the ones pictured here. We assume that in addition to these types of patterns, the reader has had many experiences with geometry through handling, measuring, and drawing. We now look more carefully at the geometrical ideas that evolve from these earlier physical experiences. We begin with an investigation which uses a small geoboard (or geopaper) to explore a small collection of points, lines, and polygons.

(c)

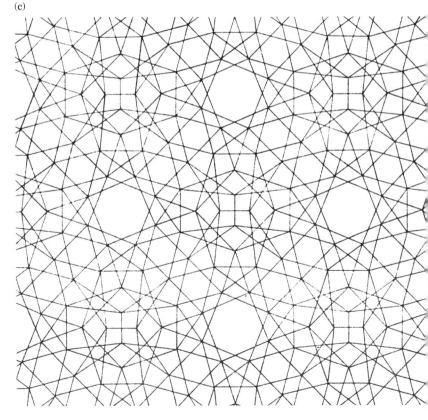

(a) A spiral pattern of prime numbers from 1 to 10,000. (b) Patterns of all lines connecting 24 evenly spaced points on a circle. (c) An Alltair Design created by Ensor Holiday based on his investigations of the architecture and pattern development in the Middle East. (Originally published by Longman Group Limited, Essex, England.)

Investigation 2.1

A. How many different length segments can you find on a geoboard with 3 nails on a side?

B. How many pairs of segments "cross" in the interior of the small shaded square?

C. How many different shaped triangles can you find on this size geoboard? (Note that triangles such as these are the same shape and would be counted only once.)

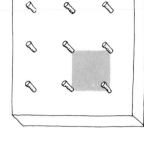

POINTS

The idea of "point" as it is used in geometry is an abstraction, and it evolves out of experiences with physical objects. In the beginning, objects such as the tip of a pencil, the corner of a block, or the forbidden point of a pair of scissors suggest the notion of point to the child. These important early experiences provide the basis for later abstraction of the idea of point and for a mature view of the concept of point as it is used in a geometric setting. (See Fig. 2.1.)

We often use the word "point" to describe a dot or inkspot on a piece of paper. This dot may represent any number of

According to the studies by Piaget, children are not ready to think about lines, triangles, squares, etc. as sets of points before eleven to twelve years of age. The early notions of points, lines, and planes should probably be developed by looking at natural three-dimensional objects in the child's environment and making special notice of the corners, edges, and surfaces.

A point as a part of a physical object

A point as the smallest dot you can draw

A point as an idea, or abstraction

POINT

Figure 2.1

B •

A •   R •

physical situations which suggest points, and consequently it conveys the notion of point for each of the objects represented. We say "draw" or "mark a point" when we mean draw a dot to represent a point. We use capital letters beside the dots as shown to name specific points.

Just as a number is an abstraction that cannot be seen or touched, so is a point. In a formal study of geometry no attempt is made to define this abstraction (a point) using previously defined simpler terms. Since the idea of point is so basic that it is used without definition, it is called an **undefined term**.

LINES

In most formal developments of geometry, "line," like point, is an undefined term. Sometimes a line is described as a set of points which satisfies a certain list of postulates. Examples from such a list of postulates are

1. Given any two different points, there is exactly one line containing them, and
2. Every line contains at least two points.

As was true for point, however, the notion of "line" or "line segment" can be cultivated in the mind of a child through experiences with physical objects. The edge of a piece of lumber, a string or wire held taut, or the paths of certain moving objects all suggest the idea of line segments or lines.

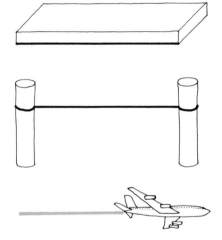

Figure 2.2

A line as part of   A line as the thinnest   A line as an idea,
a physical situation   streak you can draw   or abstraction

LINE

These physical models of lines provide the basis for later abstraction of the *idea* of line and for an abstract view of line as it is used in a formal geometric setting. (See Fig. 2.2.)

As with the word "point" and a dot on paper, we often use the word "line" to describe a streak of ink on paper. This streak may represent any number of physical situations which convey the idea of line.

Not only can the process of abstracting the idea of a line be aided by considering physical situations, it can also be expedited by thinking of the line as a special set of points, as suggested below.

Three points	. . .
Five points	. . . . .
Nine points	. . . .
More points	. . . .
More points	. . . .
More points	··························
Many more points	←————————————→
	Line

Note that a line is thought of as extending continuously in each direction. We say "draw a line" when we mean draw a mark with arrows on the end to represent a line. We name a line with a lower case letter, or by selecting two specific points and denoting them by capital letters.

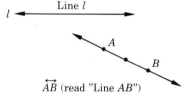

Line *l*

$\overleftrightarrow{AB}$ (read "Line *AB*")

OTHER BASIC GEOMETRIC IDEAS

In order to investigate further the basic ideas of geometry, it is helpful to review briefly some undefined terms, undefined relations, and definitions.

Just as *point* and *line* are undefined terms which are given meaning by considering physical objects, so is the idea of a "plane". A table top, or a part of a house roof might suggest this idea. A plane may also be thought of as a set of points. In addition, "space" is considered an undefined term, and we assume that experience in a three-dimensional world has helped the reader develop an intuitive notion of this idea.

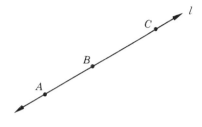

Since a line is thought of as a set of points, we can focus on selected points and mark larger dots for them on the ink streak that represents the line. In statements about points and lines, we use certain undefined relations such as "is on," "contains," or "is between."

For example, we say that "point *A is on* line *l*" to describe the situation in the figure shown here. When thinking of line *l* as a set of points, we might say "point *A belongs to* line *l*" or "point *A is an element of* line *l*" to describe the same situation. We can also say "line *l contains* point *A*" or "line *l is on* point *A*." If *A*, *B*, and *C* are points on line *l*, as shown in the figure above, we say "point *B is between A* and *C*." If *A*, *B*, and *C* are not all on the same line, we do not use "between" to describe their relationship.

Using these basic undefined notions, we can define other relations between points and lines. These definitions are given and illustrated below.

 A second-grade child was asked to draw and define a circle. Her paper looked like the following figure.

While the definitions in this section are important for upper-level students and teachers to know, extensive experiences with physical objects are much more important than definitions when helping children develop concepts. Even with secondary students, experience with physical objects is often a necessary prelude to formal definitions of basic concepts.

Definition	*Illustration*
Three or more points are **collinear** if they lie on the same line.	*A*, *B*, and *D* are *collinear*. *A*, *B*, and *C* are not collinear.
Two lines **intersect** if there is a point which is on both lines.	Line *k intersects* line *l*. Line *m* doesn't intersect line *l*.

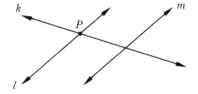

Definition	*Illustration*	

Two lines in a plane are
parallel if there is no
point which is on both
lines.

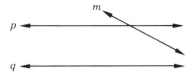

Line *p* is *parallel* to line *q*.
Line *m* is not parallel to line *q*.

Two or more lines are
concurrent if there is one
point that is on each line.

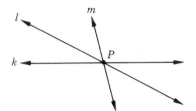

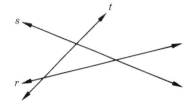

Lines *k*, *l*, and *m* are *concurrent*.
Lines *r*, *s*, and *t* are not concurrent.

Certain subsets of a single line are so useful in geometry
that they are given names. These subsets are defined, illustrated, and symbolized below.

Definition	*Illustration*	*Symbol*
A **segment** is a subset of a line which contains two points (called endpoints) and all points between them.		$\overline{AB}$
A **ray** is a subset of a line which contains a point (called the endpoint) and all points of the line on one side of this point.		$\overrightarrow{CD}$

Certain subsets of a pair of intersecting lines are also useful in geometry and are given special names. One of these subsets is defined, illustrated, and symbolized below.

Definition	*Illustration*	*Symbol*
An **angle** is the union of two rays with a common endpoint. The common endpoint is called the **vertex**. The rays are called the **sides** of the angle.	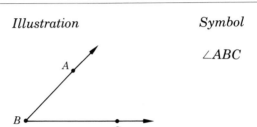	$\angle ABC$

Two cardboard strips joined with a paper fastener or two straws joined by a pipe cleaner are models helpful in introducing the concept of angle to children. These models can be placed on objects in the classroom to exhibit angles found in these physical objects.

An angle possesses a property which we might think of intuitively as "size." To describe this property we associate with each angle a number called the measure of the angle. The usual convention associates the number 180 with an angle with rays lying on a line. Consequently all other angles, such as $\angle COB$ in Fig. 2.3, possess a measure, denoted by "$m(\angle COB)$", between 0 and 180. In particular, if, in Fig. 2.3, $m(\angle AOB) = 180$, there exists a ray OE such that $m(AOE) = 90 = m(\angle EOB)$. (In a formal development of geometry, these properties of the measure of an angle are provided for through postulates.) An angle with measure 90° is called a **right angle** and the lines $\overleftrightarrow{AB}$ and $\overleftrightarrow{OE}$ in Fig.

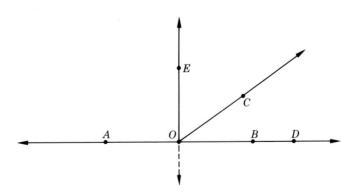

Figure 2.3 $m(\angle DOB) = 0$ $0 < m(\angle COB) < 180$ $m(\angle AOB) = 180$

2.3 are called **perpendicular** lines. If the measure of an angle is less than 90° it is called an **acute** angle, and if the measure is greater than 90° it is called an **obtuse** angle. The usual device for measuring an angle is the protractor, which is used as shown in Fig. 2.4.

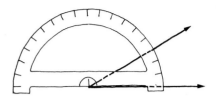

Figure 2.4

An effective way to make the transition from the physical to the abstract when discussing angles and perpendicularity is through paper folding, as illustrated in the two sequences in Fig. 2.5.

Paper folding is an effective means for giving the student "the feel" for a geometric idea. The book *Paper Folding for the Mathematics Class* (NCTM, 1957) by Donovan Johnson, is a good paper-folding resource book for grades 6 through 12.

Figure 2.5

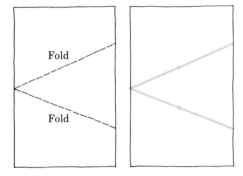

Make two folds meeting at an edge.

The two crease lines form an angle.

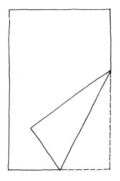

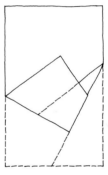

Fold one corner and crease.

Fold the crease onto itself.

The two crease lines form a right angle.

EXERCISE SET 2.1

1. Which geometric concepts are suggested by these physical situations?

   a) a taut string, held tightly at each end

   b) a beam of light from a strong searchlight

   c) a plastic hoop

   d) the blades of an open pair of scissors

   e) a fine wire bent in one place

   f) railroad tracks   g) a city intersection

2. Name at least two physical objects, not previously mentioned in this chapter, that suggest these geometric ideas.

   a) point   b) line segment
   c) plane   d) parallel lines
   e) intersecting lines   f) perpendicular lines
   g) an angle   h) collinear points
   i) concurrent lines

3. In a postulational development of Euclidean geometry, the following **incidence postulates** are accepted as true.

   P_1 For each two points, there is exactly one line containing them.

   P_2 Every line is a set of points containing at least two points.

   P_3 There are at least three noncollinear points.

   P_4 For each three noncollinear points, there is exactly one plane containing them.

   P_5 Every plane is a set of points containing at least three noncollinear points.

   P_6 If two points lie in a plane, then the line containing them lies in the plane.

   P_7 If two planes intersect, then their intersection is a line.

   P_8 There are at least four non-coplanar points.

   a) Draw a picture illustrating each of the postulates.

   b) For which of the postulates can you find a physical situation to illustrate the idea?

4. Show that a mathematical system which satisfies the eight incidence postulates of Exercise 3 possesses

   a) at least four distinct planes,

   b) at least six distinct lines.

5. Use an illustration to give meaning to the following:

   a) **midpoint** of a segment

   b) angles **vertical** to each other

   c) angles **complementary** to each other

d) angles **supplementary** to each other

e) **bisector** of an angle

6. The following situation suggests a theorem about lines and points which was discovered by Pappus in A.D. 300.

   Let p and q be any two lines. (1) Choose points A, B, C, D, E, and F alternately on p and q in any position. Then consider the pairs of lines $\overleftrightarrow{AB}$ and $\overleftrightarrow{DE}$, $\overleftrightarrow{BC}$ and $\overleftrightarrow{EF}$, and $\overleftrightarrow{CD}$ and $\overleftrightarrow{AF}$

   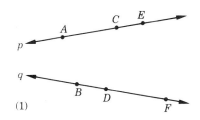
   (1)

   a) What discovery can you make about the points of intersection of these lines?

   b) Try this again using another pair of lines. Does your discovery hold true?

7. Consider three concurrent lines r, s, and t. (2) Choose any point A on line r, any point B on line s, any point C on line t. Then choose another point A' on line r, another point B' on line s; and another point C' on line t. Whenever $\overleftrightarrow{AB}$ and $\overleftrightarrow{A'B'}$, $\overleftrightarrow{BC}$ and $\overleftrightarrow{B'C'}$, $\overleftrightarrow{CA}$ and $\overleftrightarrow{C'A'}$ intersect in points P, Q, and R, what generalization can you make? Draw some pictures in which these lines do intersect to help you form the generalization.

   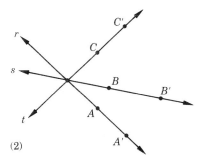
   (2)

8. Draw an accurate circle with a compass. (3) Start at a point A outside the circle and draw 6 segments ($\overline{AB}$, $\overline{BC}$, $\overline{CD}$, $\overline{DE}$, $\overline{EF}$, and $\overline{FA}$) so that each segment is tangent to the circle (intersects it in just one point), and so that the last segment ends at A. Draw another circle and do this again in a different way. Can you form a generalization about lines $\overleftrightarrow{AD}$, $\overleftrightarrow{EC}$, and $\overleftrightarrow{CF}$?

   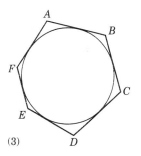
   (3)

9. Choose any six points on a circle, and label them in any order, using the letters P, Q, R, S, T, and U. (4) Connect the points in order. Label the intersection of $\overleftrightarrow{PQ}$ and $\overleftrightarrow{ST}$, the intersection of $\overleftrightarrow{QR}$ and $\overleftrightarrow{TU}$, and the intersection of $\overleftrightarrow{RS}$ and $\overleftrightarrow{UP}$ as B, C, and A respectively.

   Can you discover anything about points A, B, and C? Try this with other arrangements of six points on a circle.

   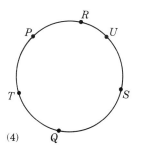
   (4)

POLYGONS

A chinese proverb says, "I hear and I forget. I see and I remember. I do and I understand." Piaget says, "motor activity is of enormous importance for the understanding of spatial thinking." Thus, it is important for students to have many physical experiences with geometric objects. The ideas of polygons can be enhanced by using cardboard cutouts or the sets of Pattern Blocks, available from McGraw-Hill Co.

In the interesting book *Flatland*, Edwin A. Abbott describes a "civilization" in which the inhabitants are restricted to the two-dimensional world of the plane. Our adaptation of his description is as follows:

> Our middle class consists of equilateral triangles. Our professional men are squares and pentagons. Next above these comes the nobility, of whom there are several degrees, beginning at hexagons and from thence rising in the number of their sides till they receive the honorable title of polygonal, or many-sided. When the number of the sides becomes so numerous, and the sides themselves so small that the figure cannot be distinguished from a circle, he is included in the circular or priestly order; and this is the highest class of all.

Although the "personalities" we depict may not be quite like those in "Flatland," the following investigation will serve to introduce the idea of a polygon.

Investigation 2.2

The picture shows a 10-sided figure (no side crossing another side) enclosing a region on a 4×4 geoboard.

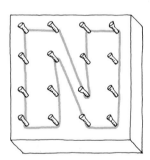

A. The largest possible number of sides for such a figure on a 4×4 geoboard is 16. Show this figure on dot paper.

B. It has been proved that for $n \geq 7$ a polygon with n^2 sides can be formed on an $n \times n$ geoboard. As suggested in part A, this is also true for $n = 4$.

   What are the largest number of sides for such a figure on a 3×3 geoboard? a 5×5 geoboard?

A definition of a simple polygon can now be given:

> Let P_1, P_2, . . . P_n be a set of n distinct points in a plane where $n > 2$. The union of the segments P_1P_2,

$P_2P_3,$ . . . P_nP_1 is a **simple polygon** provided the segments intersect only at their endpoints and no two segments with a common endpoint are collinear.

The points P_1, P_2, . . . P_n are called the **vertices** of the polygon and the segments P_1P_2, P_2P_3, . . . P_nP_1 are called the **sides** of the polygon. A segment (other than a side) joining two vertices of a polygon is called a **diagonal** of the polygon. The union of two consecutive sides of the polygon is called an **angle of the polygon.**

If the segments described in the definition intersect at a point or points in addition to their endpoints, the polygon is called a **nonsimple polygon**. These two types of polygons are illustrated in Fig. 2.6.

Figure 2.7 illustrates a way of classifying simple polygons according to whether they are convex or nonconvex.

A **convex** simple polygon satisfies the property that for each pair of points P and Q inside the polygon the segment PQ lies inside the polygon. In the case where there exists at least one pair of points P, Q such that $\overline{PQ}$ does not lie inside the polygon, the polygon is called **nonconvex.**

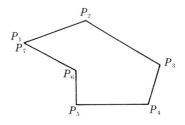

A simple polygon

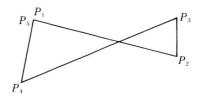

A nonsimple polygon

Figure 2.6

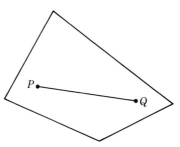

Figure 2.7   A convex polygon

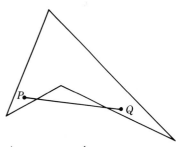

A nonconvex polygon

Polygons are named according to the number of sides they have, as indicated in Fig. 2.8. Polygons with 1 through 10, and 12, sides are named using Greek prefixes which indicate the number of sides. A polygon with 11, or more than 12, sides is referred to as an n-gon, where n is the number of sides of the polygon.

Primary children can learn to recognize different polygons by feel. The teacher might have several different polygon shapes in a bag and ask the child to reach into the bag, pick one up, and describe it exclusively by feeling it inside the bag.

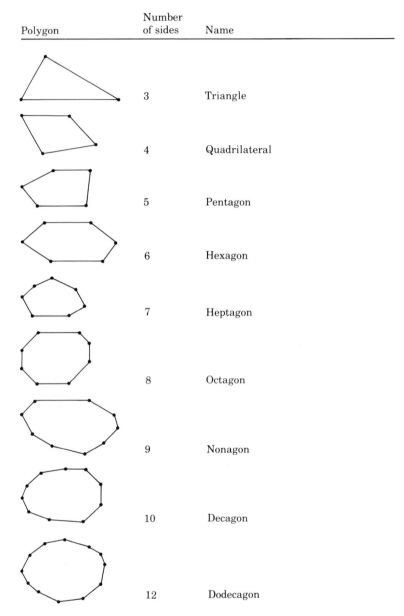

Polygon	Number of sides	Name
	3	Triangle
	4	Quadrilateral
	5	Pentagon
	6	Hexagon
	7	Heptagon
	8	Octagon
	9	Nonagon
	10	Decagon
	12	Dodecagon

Figure 2.8

In the sections that follow we will study ideas of symmetry of polygons, analyze some elementary polygons, introduce regular polygons, and explore the interesting world of the star polygons.

SYMMETRY OF POLYGONS

Each of the designs in Fig. 2.9 appears to have a certain order, balance, or beauty which makes it interesting to observe. Several of the designs possess what is known as reflectional symmetry. Some possess rotational symmetry. While a more formal definition of symmetry will be given in a later chapter, a description for both reflectional and rotational symmetry that might be appropriate for helping students learn these ideas will be presented in this section.

The set of Mirror Cards (McGraw–Hill) can help students develop a concept of reflectional symmetry. The advanced sets of these cards are appropriate for secondary-level students. Some students enjoy making their own mirror cards.

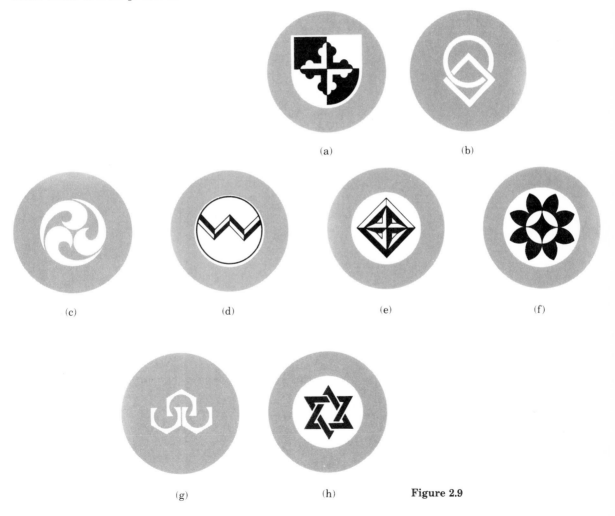

(a)

(b)

(c)

(d)

(e)

(f)

(g)

(h)

Figure 2.9

Simple ideas of symmetry can be experienced through cutting valentines and snowflakes. There are many interesting ideas for teachers in *Introducing Symmetry,* published by the University of Minnesota, which is a part of the Minnesota Mathematics and Science Teaching Project.

To test a geometric figure for reflectional symmetry, one might use what we will call (1) the "fold test" or (2) the "mirror test." To implement the fold test, trace the figure on a piece of paper. Check to see if the paper can be folded so that one half of the figure *exactly* coincides with the other half. If such a fold line can be found, the figure is said to possess **reflectional symmetry**. The fold line is called the **line of symmetry.**

To implement the mirror test, try to place a mirror or piece of plexiglass* on the figure in such a way that half of the figure and its reflection in the mirror appear to accurately form the entire figure. If the mirror can be so placed, the figure is said to possess reflectional symmetry. The line of the mirror is the line of symmetry.

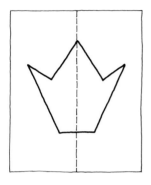

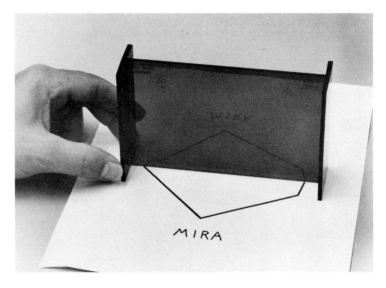

Checking for reflectional symmetry – the "mirror test."

Checking for reflectional symmetry – the "fold test."

In Fig. 2.10 we see examples of polygons with and without lines of symmetry. Use one of the tests described above to check each figure.

---

* The piece of plexiglass pictured here is commercially produced, and is called a Mira.

Figure 2.10

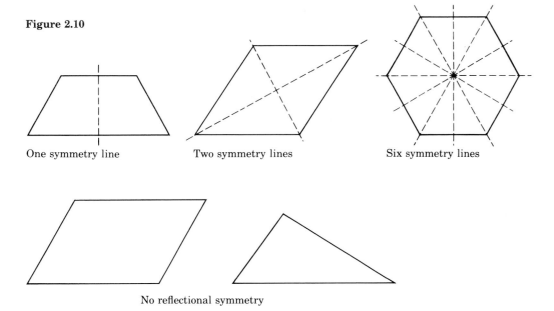

One symmetry line   Two symmetry lines   Six symmetry lines

No reflectional symmetry

To test a figure for rotational symmetry one might use what we call the "trace and turn" test. To implement this test, trace the figure on a piece of tracing paper or a piece of plastic and place the tracing directly on top of the original figure. Then, while holding one point fixed, turn the tracing until the tracing and the original again coincide, as shown in Fig. 2.11.

In Fig. 2.11, we see that after the tracing is rotated counterclockwise about the center point through an angle of 45°,

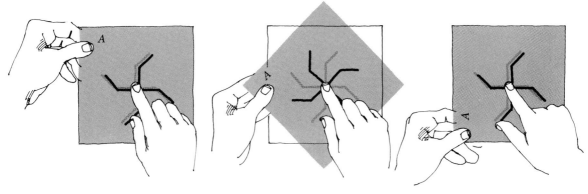

Figure 2.11 Checking for rotational symmetry—the "trace and turn test."

the figure and its tracing do not coincide. However, after a counterclockwise turn about the center point through 90°, the figure and its tracing do coincide (the tracing fits exactly upon the original figure). Thus we say the figure has 90° **rotational symmetry**. The point which is held fixed is called the **center** of rotational symmetry.

In Fig. 2.12 we see a rectangle with 180° rotational symmetry about point M and a pentagon with 72° rotational symmetry about point N. Use the trace and turn test and your protractor to check this statement.

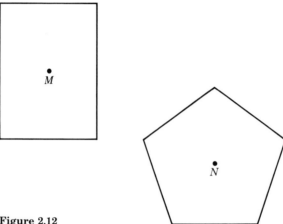

Figure 2.12

Since the tracing of the pentagon can also be made to fit exactly upon the original after rotations through 144°, 216°, and 288°, it is said to have 144° rotational symmetry, 216° rotational symmetry, and 288° rotational symmetry. Since a tracing of any figure fits upon the original figure after a rotation of 360°, we do not consider a complete rotation when describing rotational symmetry.

EXERCISE SET 2.2

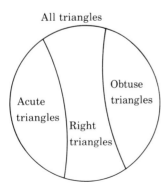

1. This Venn diagram classifies triangles according to the size of the largest type of angle occurring in the triangle.

Sketch a triangle representative of those in each region of the diagram.

2. This Venn diagram classifies triangles according to the lengths of the sides of the triangle. Note that an equilateral triangle is classified as a special type of isosceles triangle.

   Sketch a triangle representative of those in each region of the diagram.

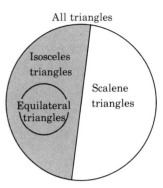

3. Examples of square dot paper and equilateral triangular (isometric) dot paper are shown here. By drawing only straight lines from one point to another, which of the triangles in Exercises 1 and 2 can you draw on each type of paper?

Isometric dot paper

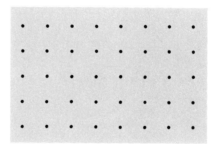

Square dot paper

4. For each part below sketch a triangle, if it is possible, satisfying the conditions that it possesses.
   a) no symmetry lines
   b) exactly one symmetry line
   c) exactly two symmetry lines
   d) exactly three symmetry lines
   e) more than three symmetry lines
   f) reflectional symmetry but no rotational symmetry
   g) rotational symmetry, but no reflectional symmetry

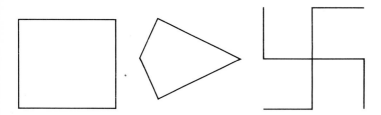

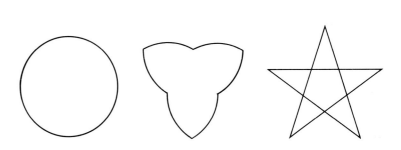

5. For each figure above sketch all lines of symmetry and describe all rotational symmetries.

*6. Four of the eight designs on page 39 possess symmetry. One of these four has only reflectional symmetry, two possess only rotational symmetry, and the fourth has both rotational and reflectional symmetry. Find these four designs. [Hint: Note that regions of different colors must be considered as different in determining the symmetry properties.]

*7. A triangle has either three lines of symmetry, one line of symmetry, or zero lines of symmetry. What are the possible number of lines of symmetry for

a) pentagons?

b) hexagons?

c) octagons?

_____

* The starred exercises are either more difficult or less central to the main topic and might be omitted in a minimal course.

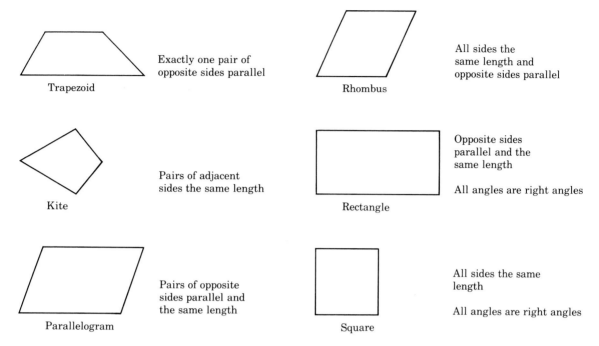

8. A **quadrilateral** is a polygon with four sides. A representative of each major type of quadrilateral is pictured above along with descriptive conditions.

   a) Describe all the lines of symmetry for each quadrilateral above.

   b) Describe all rotational symmetries for each quadrilateral above.

   c) List as many other properties of each of these quadrilaterals as you can. Consider angles, diagonals, midpoints of sides, etc.

9. Sketch, if it is possible, a quadrilateral other than those in Exercise 8

   a) which has a line of symmetry but no rotational symmetry,

   b) which has rotational symmetry but no line of symmetry.

   You need not restrict your consideration to convex quadrilaterals.

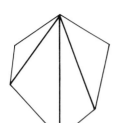

10. When the maximum number of nonintersecting diagonals is drawn, the polygon is divided up into triangles. Complete the table below and find a formula which enables you to find the number of triangles when you know the number of sides.

Number of sides of polygon	Maximum number of nonintersecting diagonals	Number of resulting triangles
3	0	1
4	1	
5		
6		
7		
.		
.		
.		
n		

Number of sides	Sum of angle measures
3	180°
4	
5	
6	
7	
.	
.	
.	
n	

11. Use the results of Exercise 10 to complete the table on the left. Search for a formula relating the number of sides of a polygon and the sum of the angle measures for the vertex angles of the polygon.

12. A pentominoe is a polygon made from five connected squares. Each square can touch another only on a complete side.

    a) Find all twelve pentominoes and show them on a geoboard or on dot paper.

    b) Which pentominoes have 90° rotational symmetry about some point? 180° rotational symmetry?

    c) Which pentominoes have exactly one line of reflectional symmetry?

    d) Which pentominoes have more than one line of reflectional symmetry?

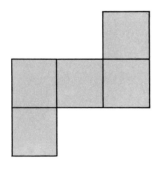

*13. Sketch a Venn diagram showing the interrelationships among the categories of quadrilaterals in Exercise 8.

14. If you had no protractor, how would you convince someone that the sum of the degree measures of the angles of
a triangle is 180°?

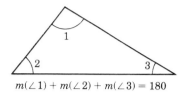

$m(\angle 1) + m(\angle 2) + m(\angle 3) = 180$

15. What is the sum of the degree measures of the vertex
angles of a quadrilateral? How could you convince someone without a formal proof that this is true?

*16. Use some cardboard strips (with punched holes and brass
fasteners) to explore the following questions:

a) Which of these figures are rigid? (i.e., which retain the
same shape without collapsing or changing to a different shape when pressure is applied?)

b) For each figure below that is not rigid, what is the
fewest number of strips that is required to make it
rigid?

c) Can you formulate a general rule for the number of
strips required to make an n-gon rigid?

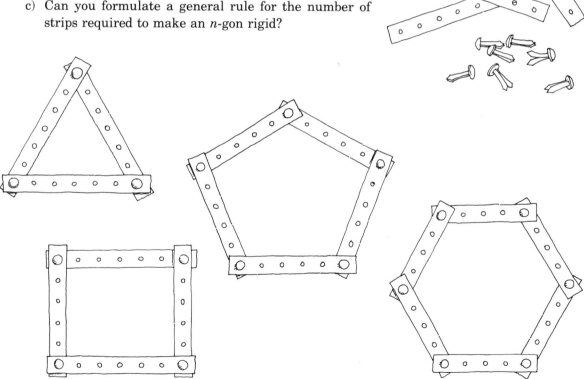

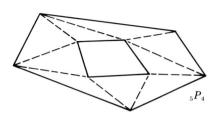

$_5P_4$

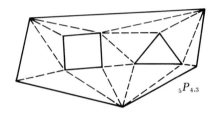

$_5P_{4,3}$

*17. A **pierced polygon** is the union of two disjoint polygons, one in the interior of the other. We use the notation $_nP_m$ to denote a pierced polygon with an m-gon interior to an n-gon.

Suppose a pierced polygonal region is divided into triangular regions. Find the relationship describing the number of triangles resulting from a pierced polygon $_nP_m$.

*18. Let $_nP_{p,q}$ represent an n-gon pierced by a p-gon and a q-gon. If the polygon $_nP_{p,q}$ is divided into triangles, what is the relationship between the numbers n, p, and q?

*19. A **parpolygon** is a polygon with an even number of sides in which opposite sides are equal and parallel. Every parpolygon with $2n$ sides can be dissected into $\dfrac{n(n-1)}{2}$ parallelograms.

   a) Show (by drawing pictures) examples of this theorem for $n = 2$, $n = 3$, and $n = 4$.

   b) Describe the dissection when the parpolygon is equilateral.

The concept of a regular polygon can be introduced nicely using geostrips or homemade cardboard strips. A pentagon, for example, can be made by fastening 5 strips together. Since a variety of different pentagons can be demonstrated with this model, it becomes clear to the student that not only side length, but also angle size should be considered when describing a regular polygon.

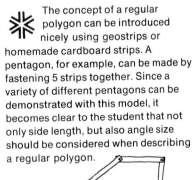

REGULAR POLYGONS

There are interesting patterns of segments, like those in Fig. 2.13 which can be constructed on a circle geoboard but which cannot be constructed on a square or triangular geoboard.

Polygons similar to these are the subject of Investigation 2.3.

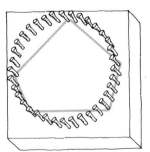
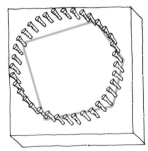

Figure 2.13

Investigation 2.3

A. On a circle geoboard (or dot paper) with 36 nails, how many different polygons with sides of equal length can you construct?

B. Can you find any relationship between the number of nails (36) and the number of sides of polygons constructed in part A?

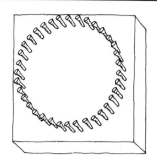

The polygons constructed in this investigation are known as regular polygons. A **regular polygon** is a simple polygon with all sides the same length and all vertex angles the same measure. A regular polygon with n sides is called a regular n-gon. Three examples of regular polygons are illustrated in Fig. 2.14.

Figure 2.14

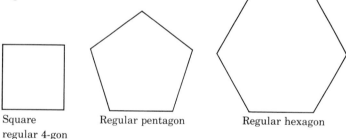

Square   Regular pentagon   Regular hexagon
regular 4-gon

It can be proven that a polygon with equal sides when inscribed in a circle (like those constructed in Investigation 2.3) will also have equal vertex angles. To further illustrate the importance of equal sides *and* equal angles in the definition of regular polygons we observe that there are polygons with all sides the same length, which do not have vertex angles of equal measure and consequently are not regular polygons. The rhombus and pentagon shown in Fig. 2.15 have all sides the same length. Neither, however, is a regular polygon. Why?

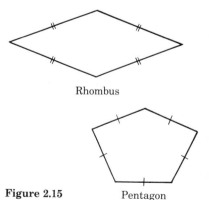

Rhombus

Figure 2.15   Pentagon

Figure 2.16

Regular polygons	Number of sides n	Diagonals divide into how many triangles?	Total number of degrees in all the triangles (i.e., the sum of measures of the vertex angles)	Number of degrees in *one* vertex angle
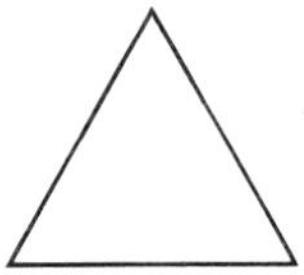 Equilateral triangle	3	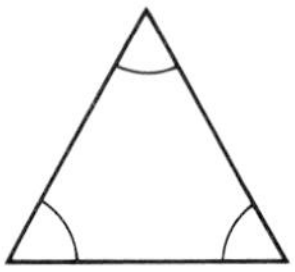 1	1(180)	$\dfrac{1(180)}{3} = 60°$
Square	4	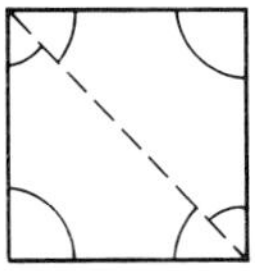 2	2(180)	$\dfrac{2(180)}{4} = 90°$
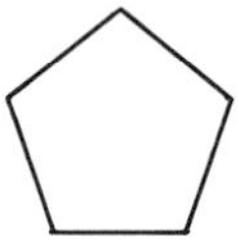 Pentagon	5	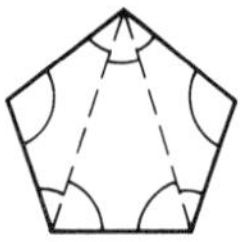 3	3(180)	$\dfrac{3(180)}{5} = 108°$
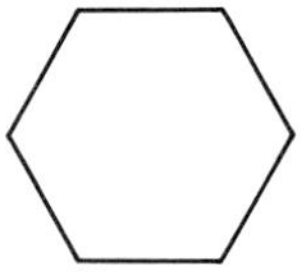 Hexagon	6	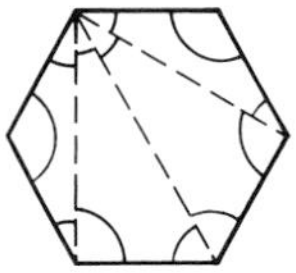 4	4(180)	$\dfrac{4(180)}{6} = 120°$
$\vdots$	$\vdots$	$\vdots$	$\vdots$	$\vdots$
	n	$n-2$	$(n-2)(180)$	$\dfrac{(n-2)180}{n}$

Similarly, there are polygons which have all vertex angles of equal measure, but which are not regular polygons. The rectangle is a simple example of such a polygon. Why is the rectangle not a regular polygon?

An important question concerning regular polygons is: *How many degrees are there in each vertex angle of a regular n-gon?* To answer this question, the reader should review Exercises 10 through 15 in Exercise Set 2.2 and study Fig. 2.16. We conclude from this table that there are $\dfrac{180(n-2)}{n}$ degrees in each vertex angle of a regular n-gon. Sometimes we may find it convenient to write this number in the equivalent form $180\left(1-\dfrac{2}{n}\right)$.

An angle with vertex at the center of a regular polygon and sides which contain adjacent vertices of the polygon is called a **central angle** of a polygon. Thus $\angle AOB$ is a central angle of pentagon $ABCDE$. A central angle of a regular n-gon has measure $360°/n$.

An angle formed by one side of a polygon and the extension of an adjacent side is called an **exterior angle** of the polygon. Thus $\angle ABF$ is an exterior angle of pentagon $ABCDE$. An exterior angle of a regular n-gon has measure $360°/n$.

CONSTRUCTION OF REGULAR POLYGONS

To draw a regular n-gon, it is convenient to begin with a circle and mark n equally spaced points on the circle. Two vertices of the n-gon are found by using a protractor to measure a central angle of the n-gon. A compass is then used to mark the remaining vertices on the circle. These vertices are connected to form the n-gon. This procedure is illustrated for the pentagon in Fig. 2.17.

While this method is appropriate, and is usually accurate enough for most practical purposes, the processes of constructing regular polygons using *only* a compass and an unmarked straight edge have long been of historical and theoretical interest. The following investigation will help answer the question, *Can each regular n-gon be constructed using only a compass and a straight edge?*

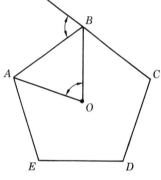

While some activities using the compass and the straight edge have been utilized with second-grade children (*Suppes' Geometric Constructions for the Primary Grades,* Singer, 1960), it is generally believed that the motor skills of the second grader are not well enough developed for extensive handling of the compass. Upper grade and junior high school students enjoy some compass constructions and the making of designs using the ruler and compass. The problems of the constructability of the polygons are sometimes motivational for secondary school students.

Figure 2.17

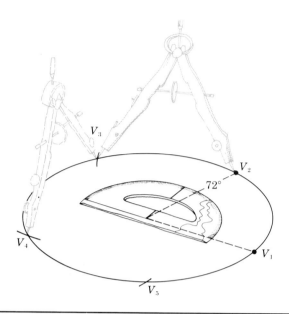

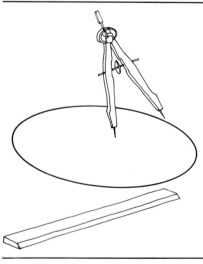

Investigation 2.4

How many regular polygons can you construct using only a compass and an unmarked straight edge?

Start with a circle. When you have constructed a certain n-gon, you may list the other n-gons that easily constructible using that n-gon.

Note that the basic ruler-compass constructions are reviewed in Appendix I.

As you may have discovered in the investigation, the regular pentagon is the first polygon which is not so easily constructed using a straight edge and compass. To construct a regular 5-gon, first draw a circle with center O (Fig. 2.18). Then proceed as follows to find on the circle the vertices

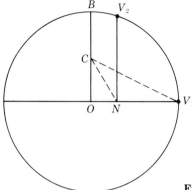

Figure 2.18

V_1, V_2, V_3, V_4, V_5 of the regular pentagon.

1. Label any point on the circle V_1 and draw $\overline{OB}$ perpendicular to $\overline{OV_1}$.
2. Join V_1 to C, the midpoint of $\overline{OB}$.
3. Bisect angle OCV_1 to obtain the point N, on $\overline{OV_1}$.
4. Construct the perpendicular to OV_1 at N and obtain the point V_2.

The segment V_1V_2 is one side of a regular pentagon and from it points V_3, V_4, and V_5 can be found. These points are then connected to complete the construction.

Once a method of constructing a regular 3-gon, a regular 4-gon, and a regular 5-gon is available, other n-gons can be constructed by successively bisecting the sides of these basic regular polygons. The following sequences describe the regular polygons which can be constructed in this way.

i) 3, 6, 12, 24, 48, . . . $3 \cdot 2^{n-1}$
ii) 4, 8, 16, 32, 64, . . . $4 \cdot 2^{n-1}$
iii) 5, 10, 20, 40, 80, . . . $5 \cdot 2^{n-1}$

Beyond this, the n-gons which can be constructed depend upon the number n. For example, a regular 5-gon can be constructed but a regular 7-gon cannot be constructed using these tools. The nineteenth-century mathematician Carl Friedrich Gauss, using the notion of Fermat Primes, completely answered this question by proving the following theorem.

Theorem. *A regular n-gon can be constructed with compass and straight edge if and only if the odd factors of n are distinct Fermat Primes.*

A Fermat Prime is a prime number which can be produced by the formula $F_k = 2^{2^k} + 1$.

The only numbers of this form known to be prime are listed below.

$$F_0 = 2^{2^0} + 1 = 2^1 + 1 = 3$$
$$F_1 = 2^{2^1} + 1 = 2^2 + 1 = 5$$
$$F_2 = 2^{2^2} + 1 = 2^4 + 1 = 17$$
$$F_3 = 2^{2^3} + 1 = 2^8 + 1 = 257$$
$$F_4 = 2^{2^4} + 1 = 2^{16} + 1 = 65537$$

Much to the surprise of Pierre de Fermat (1601–1665), after whom these primes are named, the next number generated by the formula

$$F_5 = 2^{2^5} + 1 = 2^{32} + 1 = (642) \cdot (6,700,417)$$

has more than 2 factors and hence is not a prime number. Thus, according to Gauss' theorem, a 7-gon cannot be constructed because 7 is not a Fermat Prime, a 55-gon cannot be constructed because 11 is not a Fermat Prime, and a 9-gon cannot be constructed because the odd factors of 9 are not *distinct* Fermat Primes (i.e., 3 occurs *twice* as an odd factor of 9).

EXERCISE SET 2.3

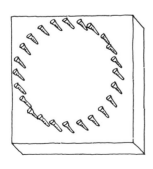

1. How many different regular polygons can you form using a circular geoboard with 24 nails?

2. Find the number of degrees in the vertex angle of a regular
   a) heptagon   b) octagon   c) nonagon
   d) decagon   e) dodecagon   f) 100-gon

3. a) Find the number of degrees in the central angle of each polygon in Exercise 2.

b) If a regular polygon with a very small central angle is chosen, what can you say about the measure of a vertex angle of this polygon?

4. Use a ruler, compass, and protractor to construct a regular heptagon.

5. Use only a compass and straight edge to construct a regular
   a) 3-gon,   b) 4-gon,   c) 6-gon,   d) 8-gon.

6. Use only a ruler and a compass to construct a regular pentagon.

7. a) How many lines of symmetry do the regular pentagon, hexagon, and octagon possess?

   b) Generalize the findings in part (a) and describe the lines of symmetry for any regular *n*-gons.

8. a) Trace a regular hexagon as in (1). Can you cut the hexagon to produce six equilateral triangles?

   b) Can you cut the hexagon to produce two isosceles trapezoids?

   c) Trace two copies of the regular hexagon. Can you cut the two hexagons to form three equilateral triangles?

9. Trace this regular hexagon (2) and cut it apart along the dotted lines. Can you place the pieces back together to form a square?

10. Use the theorem proved by Gauss and list all regular polygons with fewer than 100 sides which can be constructed with compass and straight edge.

11. Use a circle with radius 1 decimeter and construct a regular pentagon. Draw lines and label the figure as shown in the diagram. (3) Measure the lengths to the nearest millimeter and find these ratios.

   a) *AC:AB*   b) *AF:FG*   c) *FH:FG*

What do you observe?

(1)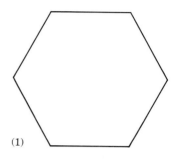

(2)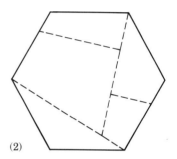

(3)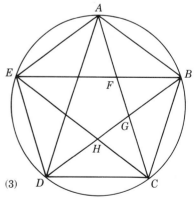

Are there any Golden Ratio relationships here?

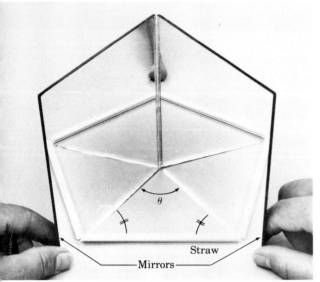

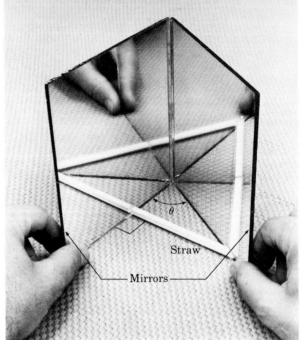

Angle between the mirrors (θ)	Number of sides of regular polygon formed
	3
	4
	5
	.
	.
	.
	n

*12. If a pair of mirrors are hinged and a straw is laid between them as shown on the left above, a regular polygon can be seen for certain values of θ (the measure of the angle formed by the two mirrors).

a) Experiment with mirrors and a straw to determine the relationship between the angle θ and the number of sides for the regular polygon formed. Complete this table.

b) Repeat the experiment in (a) with the straw positioned as shown on the right above.

*13. Cut a regular hexagon into 18 kites all the same size and shape. [*Hint*: Draw all the diagonals of the hexagon. Then draw six more short segments, strategically located.]

*14. Cut a regular hexagon into 12 rhombi all the same size and shape.

*15. Trace the three small hexagons and cut as indicated by the dotted lines. Can you place the thirteen pieces together to form the large hexagon? [*Hint*: A six-pointed star is involved.]

*16. Trace the two dodecagons and cut as indicated by the dotted lines. Can you place the thirteen pieces together to form the large dodecagon?

  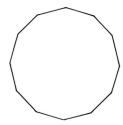

*17. The following outline describes a compass and straight-edge construction for the 17-gon. Complete the construction of a 17-gon.

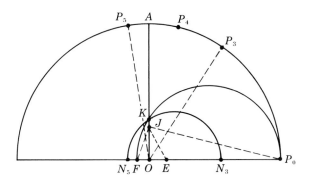

1. Construct a circle with center O and draw a diameter through O and P_0.
2. Construct $\overleftrightarrow{OA}$ perpendicular to $\overleftrightarrow{OP_0}$ and find point J one quarter of the way from O to A. Draw $\overline{P_0 J}$.
3. Label E and F on diameter $\overleftrightarrow{OP_0}$ so that $m(\angle OJE)$ is $\frac{1}{4}\, m(\angle OJP_0)$ and $m(\angle FJE)$ is $45°$.
4. Draw a circle with diameter $\overline{FP_0}$ and label the point in which that circle intersects $\overleftrightarrow{OA}$ as K.
5. Draw a circle with center E and radius EK. Label as N_3 and N_5 the points of intersection of this circle with the diameter OP_0.
6. Draw perpendiculars $\overline{N_5 P_5}$ and $\overline{N_3 P_3}$.
7. Bisect $\angle P_3 O P_5$ to obtain point P_4. Then arc $P_3 P_4$ is $\frac{1}{17}$ of the circumference of the original circle.

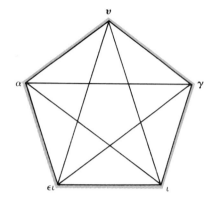

STAR POLYGONS

The five-pointed star, often called a "pentagram," was chosen as the sacred symbol of the Pythagorean Society of ancient Greeks because of its special beauty and because of the appearance of the Golden Ratio when certain of its segments were compared(see Exercise 11, Exercise Set 2.3). This symbol was worn on their clothing, and around the five points were placed the letters of the Greek word for health,"$\upsilon\gamma\iota\epsilon\iota\alpha$" (or hygeia).

This star, like the regular pentagon, can be constructed from five equally spaced points on a circle. Beginning at V_1 and going around the circle in one direction, every second point is joined by a segment. The star is a nonsimple polygon and is called a **star polygon.** Like the regular pentagon, it has five vertices and five sides. Since all five segments are of the same length and all vertex angles are of the same measure, it is called a **regular star polygon.** Since it is constructed from *five* points with every *second* point joined (i.e., $V_1V_3V_5V_2V_4V_1$ in Figure 2.19), this regular star polygon is denoted by $\left\{ \begin{matrix} 5 \\ 2 \end{matrix} \right\}$.

In Fig. 2.20 we analyze all the possibilities using five equally spaced points on a circle. Note that there is exactly one star polygon with five sides, since $\left\{ \begin{matrix} 5 \\ 2 \end{matrix} \right\}$ and $\left\{ \begin{matrix} 5 \\ 3 \end{matrix} \right\}$ represent the same set of points and hence the same star polygon.

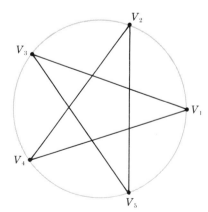

Figure 2.19

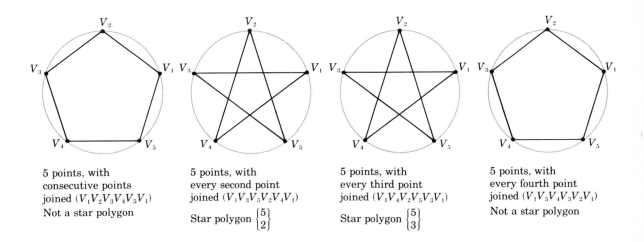

5 points, with consecutive points joined $(V_1V_2V_3V_4V_5V_1)$

Not a star polygon

5 points, with every second point joined $(V_1V_3V_5V_2V_4V_1)$

Star polygon $\begin{Bmatrix} 5 \\ 2 \end{Bmatrix}$

5 points, with every third point joined $(V_1V_4V_2V_5V_3V_1)$

Star polygon $\begin{Bmatrix} 5 \\ 3 \end{Bmatrix}$

5 points, with every fourth point joined $(V_1V_5V_4V_3V_2V_1)$

Not a star polygon

The process for constructing the star polygon $\begin{Bmatrix} 5 \\ 2 \end{Bmatrix}$ can be generalized, although we need to proceed with a degree of caution. Suppose we begin with eight equally spaced points on a circle and join every second point (Fig. 2.21). A closed path is obtained before all eight points have been reached. We do not obtain a star polygon. On the other hand, if we join every third point (Fig. 2.22), a nonsimple closed polygon with eight sides is obtained which may also be called a regular star polygon. Since we began with eight equally spaced points and joined every third point, this star polygon is denoted $\begin{Bmatrix} 8 \\ 3 \end{Bmatrix}$.

Figure 2.20

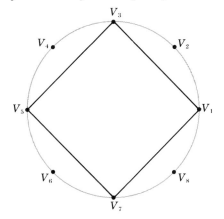

Figure 2.21

8 points, with every second point joined $(V_1V_3V_5V_7V_1)$

Not a star polygon

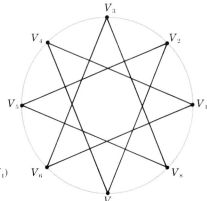

Figure 2.22

8 points with every third
point joined ($V_1V_4V_7V_2V_5V_8V_3V_6V_1$)

Star polygon $\begin{Bmatrix}8\\3\end{Bmatrix}$

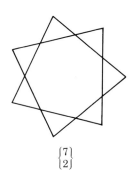

$\begin{Bmatrix}7\\2\end{Bmatrix}$

Usually, a regular star polygon which has been constructed from n equally spaced points on a circle by joining every d^{th} point is denoted $\begin{Bmatrix}n\\d\end{Bmatrix}$. We have already seen that there is no star polygon for the pair $\begin{Bmatrix}8\\2\end{Bmatrix}$, which raises the question: *Which pairs of numbers $\begin{Bmatrix}n\\d\end{Bmatrix}$ represent a regular star polygon?* (The exercises will help answer this question.) The reader should convince himself that the star polygons $\begin{Bmatrix}5\\2\end{Bmatrix}$ and $\begin{Bmatrix}8\\3\end{Bmatrix}$, together with the star polygons in Fig. 2.23, are the only star polygons with fewer than ten sides.

$\begin{Bmatrix}7\\3\end{Bmatrix}$

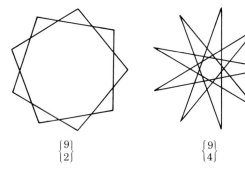

Figure 2.23

EXERCISE SET 2.4

1. Answer these questions for each star polygon in Fig. 2.23.
   a) How many sides does each polygon possess?
   b) If the vertices are labeled V_1, V_2, V_3, V_4, etc., name all the sides of each polygon.
   c) How many lines of symmetry does each possess?
   d) Describe the rotational symmetry of each polygon.

2. Using circle dot paper with eight equally spaced points, show by drawings that $\begin{Bmatrix} 8 \\ 3 \end{Bmatrix}$ and $\begin{Bmatrix} 8 \\ 5 \end{Bmatrix}$ are the only eight-sided star polygons, and that in fact they are the identically same star polygon.

3. Use circle dot paper with ten equally spaced dots to determine how many distinct ten-sided regular star polygons exist.

4. Use circle dot paper with twelve equally spaced dots to determine how many distinct twelve-sided regular star polygons exist.

5. Using the experience gained in Exercises 2, 3, and 4, can you make any guesses as to which star polygons exist? In other words, there is a numerical relationship between n and d if the star polygon $\begin{Bmatrix} n \\ d \end{Bmatrix}$ exists. What is it?

6. $\begin{Bmatrix} 5 \\ 2 \end{Bmatrix}$ and $\begin{Bmatrix} 5 \\ 3 \end{Bmatrix}$ represent the same star polygon.

   $\begin{Bmatrix} 7 \\ 3 \end{Bmatrix}$ and $\begin{Bmatrix} 7 \\ 4 \end{Bmatrix}$ represent the same star polygon.

   $\begin{Bmatrix} 8 \\ 3 \end{Bmatrix}$ and $\begin{Bmatrix} 8 \\ 5 \end{Bmatrix}$ represent the same star polygon.

   a) Which other symbol represents the star polygon $\begin{Bmatrix} 10 \\ 3 \end{Bmatrix}$?
      $\begin{Bmatrix} 11 \\ 5 \end{Bmatrix}$? $\begin{Bmatrix} 12 \\ 7 \end{Bmatrix}$?

   b) In general, the star polygon $\begin{Bmatrix} n \\ d \end{Bmatrix}$ is the same as the star polygon $\begin{Bmatrix} n \\ ? \end{Bmatrix}$. Explain why this is true.

7. **Theorem.** *The n-sided star polygon* $\left\{ {n \atop d} \right\}$ *exists if and only if* $d \neq 1$, $d \neq n - 1$, *and n and d are relatively prime.* Use this theorem to determine how many eleven-sided star polygons exist.

*8. **Theorem.** *The number of numbers less than n and relatively prime to n is* $n \left(1 - \dfrac{1}{p_1} \right) \left(1 - \dfrac{1}{p_2} \right) \ldots$ *where* p_1, p_2, . . . *are prime factors of n.*

a) How many numbers are less than 36 and relatively prime to 36?

b) How many star polygons are there with 36 sides?

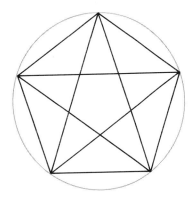

9. This figure is sometimes called the five-point "mystic rose."

a) How many different geometric figures can you find in this five-point mystic rose?

b) How many different geometric figures can you find in a six-point mystic rose?

10. Trace and cut up three copies of small star S and fit the resulting twelve pieces together to form this large six-pointed star.

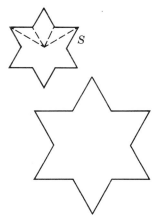

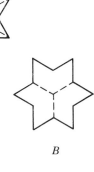

A

B

11. Trace one copy of star A and two copies of star B and cut them up as indicated. Fit the resulting twelve pieces together to form this large six-pointed star.

DISCOVERING THEOREMS ABOUT POLYGONS

Many interesting and unexpected relationships for polygons have been discovered by mathematicians over the course of many centuries. Relationships so regular and so surprising as to be considered beautiful have emerged from figures as simple as a triangle or as irregular as a general quadrilateral. In this section we shall explore a few such relationships involving triangles and quadrilaterals. To allow the reader to discover for himself some of these beautiful relationships, much of the material of this section takes the form of investigations or exercises.

There are certain "process" goals for learning mathematics. That is, students should learn the process of active inquiry, of discovery of relationships, of formulating and testing conjectures, and of critical and analytical thinking. In many cases, the process is more important than the content. Thus, students of all ages should have as many opportunities as possible to discover geometric relationships.

We begin with an investigation which involves constructing certain lines which relate to a triangle. The reader may wish to consult Appendix A for a review of basic constructions with compass, straight edge, and Mira.

Investigation 2.5

A. Draw a large acute, scalene triangle of the general shape of △ *ABC*.

1. Draw a segment from each vertex to the midpoint of the opposite side.
2. Draw from each vertex a segment perpendicular to the opposite side.
3. Draw the lines passing through the midpoints of the sides and perpendicular to the sides of the triangle.

B. The construction in part A introduces new lines and new points (the midpoints of sides, the points of intersection of lines). What relationships have you discovered among these points and lines?

C. For each relationship mentioned in part B, check and see if the relationship remains true for other triangles (obtuse, isosceles, etc). Conjecture some theorems.

One of the functions of this investigation is to introduce the reader to several important segments and lines associated with a triangle—*medians*, *altitudes*, and *perpendicular bisectors* of the sides. These terms are defined and illustrated in Fig. 2.24.

Figure 2.24   *Definition*   *Illustration*

A **median** of a triangle is a segment from a vertex to the midpoint of the opposite side.

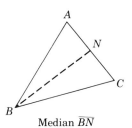
Median $\overline{BN}$

An **altitude** of a triangle is a segment from a vertex perpendicular to the line containing the opposite side. The point of intersection of this line and the altitude is called the *foot* of the altitude.

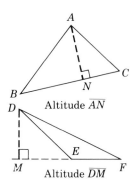
Altitude $\overline{AN}$

Altitude $\overline{DM}$

A **perpendicular bisector** of a triangle is a line perpendicular to a side and containing the midpoint of the side.

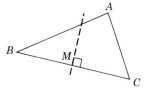
Perpendicular bisector at M

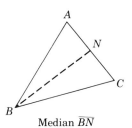
Figure 2.25

Accurate constructions in Investigation 2.5 may have led the reader to observe that the three medians are concurrent

(Fig. 2.25). Was this a coincidence, or are the medians concurrent for all triangles? The answer is emphatically stated in the following theorem, which is stated here without proof.

Theorem 1. *For any triangle, the medians are concurrent.*

At first glance this theorem seems almost expected. Perhaps the reader has learned of the theorem in a previous geometry course and is not surprised to rediscover it. However, on second glance this concurrence relation is almost astounding. Three randomly chosen lines are certainly not expected to be concurrent. Although medians are not randomly selected segments, the triangle itself was randomly chosen; the triangle was determined from three randomly selected nonconcurrent lines.

For similar reasons the next two theorems, which the reader may have discovered, are often unexpected.

Theorem 2. *For any triangle, the lines containing the altitudes are concurrent.*

Theorem 3. *For any triangle, the perpendicular bisectors are concurrent.*

These three theorems describe three points associated with each triangle. The point of intersection of the medians is the **centroid** (*G*); the point of intersection of the altitudes is the **orthocenter** (*H*); and the point of intersection of the perpendicular bisectors is the **circumcenter**(*O*).

The circumcenter is the center of a circle called the **circumscribed circle,**which contains all three vertices of a triangle (Fig. 2.26).

A theorem perhaps more stunning and beautiful than the first three is one which relates the centroid, the orthocenter, and the circumcenter. Who would expect these three points to be collinear?

Theorem 4. *In any triangle the centroid, orthocenter, and circumcenter are collinear* (Fig. 2.27).

Figure 2.26

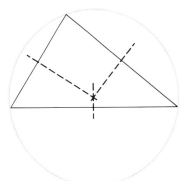

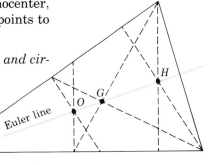

Figure 2.27

The line containing these three points is called the **Euler line**, named after the Swiss mathematician Leonard Euler who discovered this theorem. Some questions which the reader might explore are:

1. Is point G always between points O and H?

2. Is there any numerical relationship relating the distances between these three points?

3. What happens when the original triangle is isosceles? equilateral?

In each of the following exercises a setting is provided in which at least one interesting potential theorem can be discovered. The reader may find it helpful to work on these exercises with another person and together seek interesting relationships or patterns of regularity (i.e. regular polygons, concurrence, collinearity, etc.)

Here are some suggestions which might help the reader in attempting to discover interesting relationships in a given situation.

 i) Draw large figures since they make it easier to recognize relationships.

 ii) Use at least three different types of figures for each situation, to verify that the relationship discovered does not depend upon consideration of a figure of a particular size or shape.

iii) Use measurement freely, especially use a protractor and a ruler marked in millimeters.

iv) A Mira is often helpful in making constructions since fewer construction lines take less time and don't obscure the relationship.

 v) Varying the conditions of the problem may suggest a new, perhaps interesting, relationship. Often a "What if" question sparks an idea. For example, if the situation you have been exploring deals with quadrilaterals, you might ask, What if I began with a triangle or a pentagon?

EXERCISE SET 2.5

1. Begin with *any quadrilateral* and join the midpoints of adjacent sides. (1) The resulting figure is of what type?

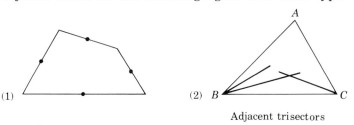

(1)   (2) B

Adjacent trisectors

2. a) Begin with *any triangle, △ABC.* (2) Use a protractor to trisect each of the angles *A, B, C.* Extend adjacent trisectors until they intersect. Connect these three points of intersection. What do you discover?

   b) Repeat part (a) by trisecting the exterior angles of △*ABC.*

3. Begin with any triangle, △*ABC.* (3) On each side of this triangle construct equilateral triangles external to the triangle (like the one shown here). Find the centroid of each of these three triangles and join them by segments. What do you discover?

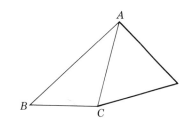

(3)

4. Begin with any triangle. Draw the three angle bisectors of the triangle.

   a) What do you discover about these three bisectors?

   b) What do these three lines have to do with the inscribed circle of a triangle, the circle which touches each side of the triangle at exactly one point?

5. Begin with any triangle, △*ABC.* (4) Find the midpoints of the sides and label them *L, M, N.*

   a) How does △*LMN* compare with each of the triangles, △*ALM,* △*CLN,* △*BNM?*

   b) For each of the three triangles containing a vertex of △*ABC* in part (a), construct the circumcenters and the orthocenters. Connect these three circumcenters. Similarly connect these three orthocenters. What do you discover?

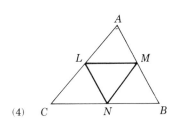

(4)

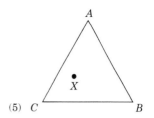

6. Choose any point (X) inside an equilateral triangle. (5) Construct the perpendiculars from X to each side of the triangle ($\overline{XE}$, $\overline{XF}$, $\overline{XG}$). Compare the length of the altitude of $\triangle ABC$ with the sum of the length of these three segments.

(5)

7. Begin with any convex quadrilateral. (6) On each side construct a square lying external to the quadrilateral. Find the centers of these squares and label them X_1, X_2, X_3, X_4. If X_1 and X_3 are centers of opposite squares, compare segments $\overline{X_1X_3}$ and $\overline{X_2X_4}$. What do you discover?

(6)

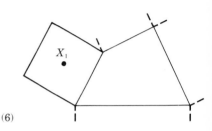

8. Begin with any parallelogram $ABCD$. (7) On each side of the parallelogram construct squares lying external to the parallelogram. The centers of these squares X_1, X_2, X_3, X_4 are the vertices of what type of figure?

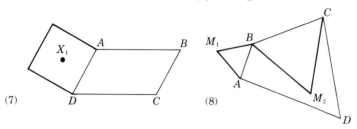

(7)   (8)

9. Begin with any convex quadrilateral $ABCD$. (8) Construct equilateral triangles ABM_1, BCM_2, CDM_3, and ADM_4 where they are alternately exterior and interior to the quadrilateral. The points M_1, M_2, M_3, M_4 are vertices of what kind of polygon?

10. Begin with any triangle, $\triangle ABC$. (9) Construct squares on sides $\overline{AB}$ and $\overline{BC}$ external to the triangle. If M is the midpoint of $\overline{AC}$, and X and Y are the centers of the constructed squares, how are these three points related?

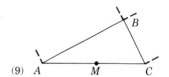

(9)

11. a) Begin with any triangle. Construct the following nine points:

    i) the midpoints of the three sides,

    ii) the feet of the three altitudes,

    iii) the midpoints of the segments from the three vertices to the orthocenter.

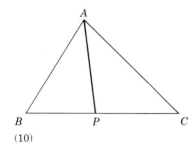

(10)

How are these nine points related?

b) Is there any relationship between the discovery found in part (a) and the Euler line?

*12. Begin with $\triangle ABC$. (10) Bisect $\angle A$. Measure (to the nearest millimeter) to find the ratio $\dfrac{BP}{PC}$.

a) Can you find another ratio in the triangle that is approximately equal to this ratio?

b) Try this with other angles in $\triangle ABC$ and with other triangles. Does it seem to hold true? Can you state a theorem?

*13. Begin with any $\triangle ABC$.(11) Bisect the external angles at A and C and the internal angle at B.

a) What do you discover?

b) Try this with other angles in the triangle and with other triangles. Does your discovery hold true? Can you state a theorem?

c) Consult a reference and find out what an *excenter* of a triangle is. How does it apply to this situation?

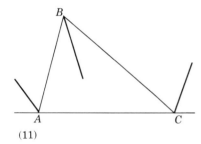

(11)

*14. Start with $\triangle ABC$. Bisect $\angle A$ and $\angle B$. Bisect the external angle at C. If the sides of $\triangle ABC$ are necessarily extended, what can you discover about the points where the three bisectors intersect the sides of the triangle opposite the bisected angles? Can you state a theorem?

*15. The incenter of a triangle is the point of intersection of the angle bisectors. Begin with $\triangle ABC$. Find the incenter and draw the inscribed circle. What can you say about the lines determined by connecting vertices to the points of tangency of the incircle? Can you state a theorem?

*16. Begin with $\triangle ABC$. Find the circumcenter and draw the circumscribed circle. Now choose *any* point on this circle and construct the perpendiculars from this point to each of the three sides of the triangle (sides may be extended if necessary). What do you discover? Do this again with a different point on the circle, with a different triangle. What do you discover?

*17. Begin with $\triangle ABC$. Choose one point, anywhere you like, on each side of the triangle. Construct three circles, each of which is determined by a vertex point and the two points nearest it. Do this again with different points and with a different $\triangle$. What do you discover?

*18. Given an n-gon $A_1A_2 \ldots A_n$, construct an n-gon $B_1 \ldots B_n$ by joining the midpoints of the edges of $A_1 \ldots A_n$. (12)

> *Conjecture:* The n-gon $B_1 \ldots B_n$ is always convex even if $A_1 \ldots A_n$ is not.

Find an example to show that this conjecture is false.

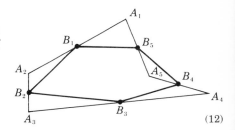

(12)

*19. Begin with $\triangle ABC$. Find an accurate way to locate by construction points P_1 (halfway around the $\triangle$ from vertex A), P_2 (halfway around the $\triangle$ from vertex B), and P_3 (halfway around the $\triangle$ from vertex C).

Consider lines AP_1, BP_2, and CP_3. What do you discover?

BIBLIOGRAPHIC REFERENCES

[6, 143–149], [9], [17, 32–35, 52–59, 107–116], [26, 26–36], [34, 36–46, 50–59], [39, 65–72], [40, 173–182], [43, 43–51, 136–146], [47], [52], [56], [66], [67], [88], [98, 152–154], [99, 46–58, 182–200]

PEDAGOGICAL ACTIVITIES FOR THE TEACHER

1. A concept has been learned when the student has a basic understanding of a mathematical idea. Concept cards are often used to help teach concepts. For example, the concept of a triangle may be taught to young children by using the ideas being suggested by this concept card.

Note that the card teaches the concept of triangle using examples along with the appropriate nonexamples. Choose at least one of the concepts presented in this chapter and develop a concept card which teaches this concept.

2. Study the teacher's manual for the pattern blocks available from McGraw-Hill. Describe at least one way these blocks could be used to help a student learn a geometric concept.

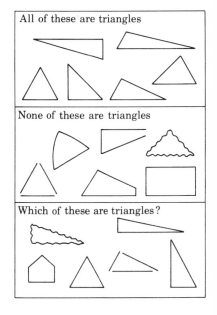

3. Articles on the use of the geoboard, appropriate for students at almost every level, are available in journals such as *The Arithmetic Teacher, The Mathematics Teacher,* and *Mathematics Teaching.* Find an appropriate article for the level of your interest and evaluate the usefulness of the ideas presented.

4. Make a set of cards which can be used for a game of "Concentration" with students. The cards should include pictures of basic polygons, definitions, and so on.

5. Plan a sequence of activities on symmetry for students. You may want to include paper folding and cutting, ink blot designs, symmetry in nature, repeating geometric figures on graph paper, and mirror activities. Include objectives for each activity and one test question with which you could ascertain whether or not the objective had been met.

6. Do posters in the classroom have value? Make a poster that would motivate students to draw star polygons and investigate their properties. Design the poster in such a way that students are encouraged to become actively involved and leave the results of their work in an envelope attached to the poster. Place the poster in a classroom and devise a means for evaluating the effectiveness and value of the poster.

7. Look in references such as the book *Recreational Problems in Geometric Dissections and How to Solve Them* by Harry Lindgren, and find at least five dissection problems involving polygons which you feel would be interesting to students at a given level. Try these puzzles with selected students and evaluate their usefulness.

8. Select a grade level and work with a group of students using the Mirror Cards published by McGraw-Hill. Ask specific questions that you have previously prepared to stimulate their interest in the activities. Find the answers to these questions: a) What is the most difficult set of Mirror Cards which these students can complete accurately? b) Can these students make Mirror Cards themselves? c) Are the Mirror Cards interesting and appropriate for these students? Following this trial of the Mirror Cards, write a set of objectives which specify the learning you hope will be affected by using the Mirror Cards.

3

Patterns of Polygons in the Plane

Figure 3.1

From *Regular Features*, L. Fejes-Toth

INTRODUCTION

In Chapter 2 we studied properties of polygons. In this chapter we explore patterns formed by combinations of polygons in the plane. These patterns are intimately related to what might be called the "art of ornamentation." Many examples of floor, wall, and fabric patterns from early Egyptian, Greek, and Chinese civilizations survive to this day. Figure 3.1 (1, 2, and 3) shows several early Greek frieze patterns; Fig. 3.1 (4, 5, 6, 7) shows Egyptian wall patterns. Figure 3.2 (1, 2) shows patterns of Chinese origin.

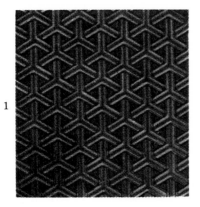

1

2

From *Regular Features*, L. Fejes-Toth

Figure 3.2

Weyl writes (1952) [107]:

One can hardly overestimate the depth of geometric imagination and inventiveness reflected in these patterns. Their construction is far from being mathematically trivial. The art of ornament contains in implicit form the oldest piece of higher mathematics known to us.

The mathematics referred to in this quotation is the theorem that there are exactly 17 two-dimensional crystallographic groups. (This theorem provides the basis for the popular statement that "there are exactly 17 essentially different wallpaper

patterns.") The first direct mathematical treatment of this theorem was given by Fedorov in 1891, yet the ornaments decorating the walls of the Alhambra in Granada indicate that all 17 types of patterns were known to the Moors.* In fact, the Moors may have provided a creative stimulus for Maurits Escher (1898–1972), the Dutch painter whose art has been a special source of enjoyment for persons interested in mathematics. In a recent book [35], Escher says:

> This is the richest source of inspiration I have ever struck; nor has it yet dried up . . . a surface can be regularly divided into, or filled up with, similar-shaped figures (congruent) which are contiguous to one another, without leaving any open spaces. The Moors were past masters of this. They decorated walls and floors, particularly in the Alhambra in Spain, by placing congruent multi-colored pieces of majolica (tiles) together without leaving any spaces between . . .

Figures 3.3 and 3.4 show some of the works of the Moors. Figure 3.5 shows this type of art coming to life in one of Escher's creations.

Today also we find many interesting ornamental patterns in our architecture, one of which appears at the left. While a thorough analysis of patterns would take us beyond the scope of this book, we shall explore a particular type of pattern called a tiling or a tessellation.

This chapter is divided into three sections. In the first section we study tessellations which can be generated by repeated use of one polygon shape. In the second section we consider tessellations formed by using more than one polygon shape, but we require that all the polygons be regular polygons. Finally, in the third section we illustrate how mirrors can be used to generate many interesting tessellations, and how it is possible to construct tessellations made of curved figures rather than polygon shapes.

---

* The Moors occupied Spain from 711 to 1492. They were forbidden by their religious credo to draw living objects.

Figure 3.3

Alhambra drawings, M. C. Escher

Figure 3.4

Alhambra drawings, M. C. Escher

Day and Night, M. C. Escher

Figure 3.5

TESSELLATIONS OF POLYGONS

A student might be asked, "In how many different ways can you tile a floor?" This question might be asked at various grade levels with differing results in each case. Oftentimes actual floor tiles can be acquired from local tile companies. Then students can actually use them to show a tiling. This is an excellent opportunity for hands-on experience with physical objects.

In this section we shall study patterns in the plane constructed using polygonal figures which completely cover the plane with no "holes" and no "overlapping." Such a pattern is called a **tessellation** or a **tiling** of the plane.

Before reading further, complete a thorough search for answers to the question asked in Investigation 3.1.

One of the first questions suggested by the investigation is, *Which regular polygons tessellate the plane?* Figure 3.6 describes three basic tessellations of the plane with regular polygons.

Investigation 3.1

Square regions can be arranged into a repeating pattern which completely covers the plane. There are no "holes" and no "overlapping" areas. We say that squares "tile," or "tessellate" the plane.

Use cardboard or plastic "polygons" to search for other quadrilaterals, triangles, or regular polygons which will tessellate the plane. Then use tracing paper to show tessellations of those figures below which will tessellate the plane.

Can you state some general conclusions about which polygons will tessellate the plane?

Parallelogram

Trapezoid

Kite

Any quadrilateral

Equilateral triangle

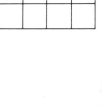
Isosceles triangle

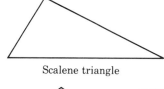
Scalene triangle

Pentagon

Hexagon

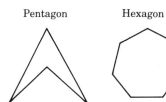
Nonconvex quadrilateral

Heptagon

Regular Tessellations

A picture of a portion of the tessellation	Description of the tessellation	Notation
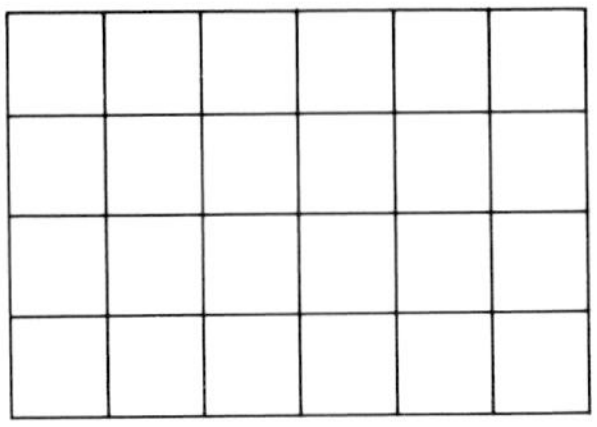	Made up of regular **4**-gons with **4** of them surrounding each vertex	4,4,4,4 or 4^4
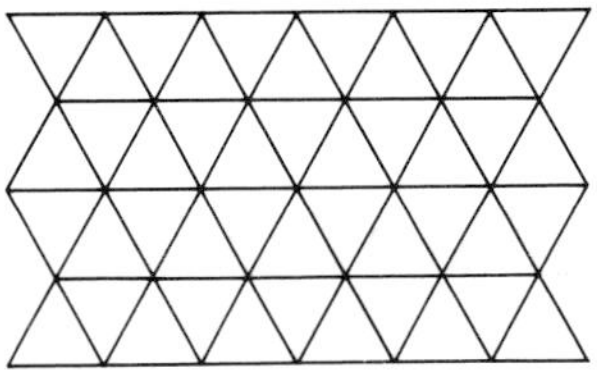	Made up of regular **3**-gons with **6** of them surrounding each vertex	3,3,3,3,3,3 or 3^6
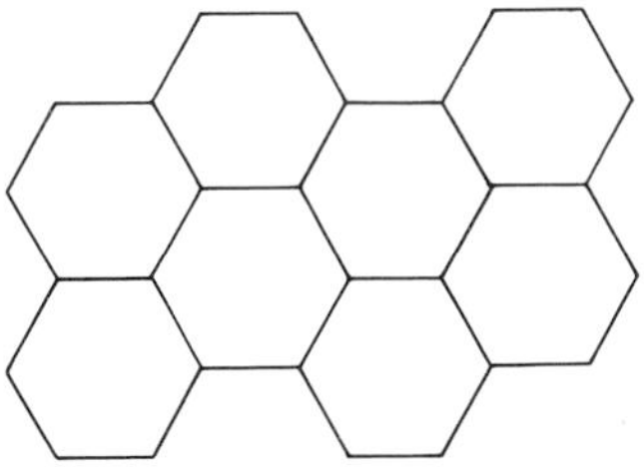	Made up of regular **6**-gons with **3** of them surrounding each vertex.	6,6,6 or 6^3

Figure 3.6

The tessellations of equilateral triangles and squares in Fig. 3.7 are different from the tessellations in Fig. 3.6 in that the vertices of the triangles and squares in Fig. 3.7 coincide with the midpoints of the sides of other squares and triangles. To explain the difference more formally we refer to what are called vertex figures. In Fig. 3.8 the polygons drawn in gray are polygons whose vertices are the midpoints of the polygonal edges emanating from a vertex of the tessellation and are known as **vertex figures**. A tessellation is a **regular tes-**

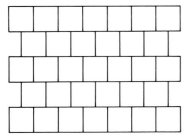
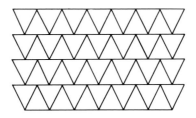

Figure 3.7

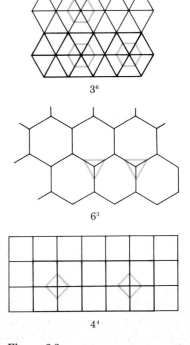

3^6

6^3

4^4

Figure 3.8

sellation if it is constructed from regular convex polygons of one size and shape such that each vertex figure is a regular polygon. It can be proven that in regular tessellations all vertex figures are congruent to each other.

Note (Fig. 3.8) that the vertex figure for the regular tessellation 3^6 is a regular hexagon; for the regular tessellation 6^3 it is an equilateral triangle, and for the regular tessellation 4^4 the vertex figure is a square.

On the other hand, we see (Fig. 3.9) that the vertex figures for the tessellations in Fig. 3.7 are isosceles triangles and trapezoids, respectively, and are not regular polygons. Consequently the tessellations in Fig. 3.9 are not regular tessellations.

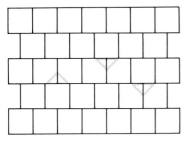
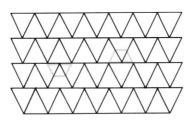

Figure 3.9

The question remains, how many regular tessellations are there in addition to those in Fig. 3.6? Considering the regular pentagon, we see (Fig. 3.10) that three regular pentagons surrounding a vertex leave a gap and four regular pentagons surrounding a vertex overlap. So there can be no regular tessellation of pentagons.

Figure 3.10

Attempts to tessellate with other regular polygons will show that if there exists a regular tessellation of p-gons, the vertex angle of the p-gons must divide 360. Furthermore, there must be at least three polygons around each vertex in a regular tessellation so the vertex angle cannot be greater than 120°. In Table 3.1 we see that the only vertex angles which are less than or equal to 120°, and which also divide 360, are the vertex angles of equilateral triangles, squares, and regular hexagons. Consequently, the tessellations 3^6, 4^4, and 6^3 pictured in Fig. 3.6 are the only regular tessellations.

Table 3.1

Number of sides of a polygon	Number of degrees in one vertex angle of the polygon
3	60°
4	90°
5	108°
6	120°
7	128⁴/₇°
8	135°
.	.
.	.
.	.
n	$\dfrac{(n-2)180}{n}$

Another question suggested by the investigation is, *Which triangles will tessellate the plane?* To answer this question, consider any triangle and note that this triangle can be paired with another triangle of the same size and shape to form a parallelogram (Fig. 3.11). Since all parallelograms tes-

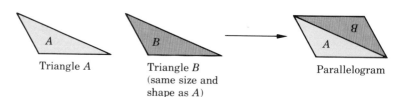

Triangle A | Triangle B (same size and shape as A) | Parallelogram

Figure 3.11

sellate the plane (the reader should use cutouts or drawings to see that this is plausible), a tessellation of the arbitrarily chosen triangle is effected by the tessellation of the parallelograms. Thus *any triangle will tessellate the plane.*

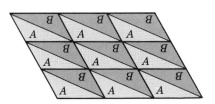

A third question suggested by the investigation is, *Which quadrilaterals will tessellate the plane?* To answer this question, consider *any* quadrilateral and note that:

i) Successive 180° rotations about the midpoints (*A, B,* and *C*) of the sides of the quadrilateral will generate four copies of the quadrilateral around a vertex (*V*). (The reader should check this with tracing paper.) In particular, each of the angles around *V* is a distinct angle of the quadrilateral, and these four angles have a sum of 360°.

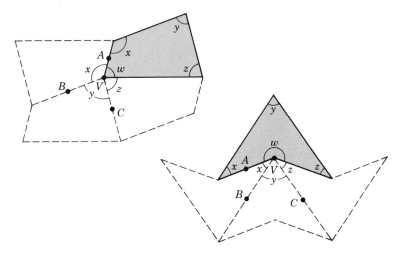

ii) Since this process can be carried out at each vertex, it is possible to generate a tessellation of the original quadrilateral.

Note that the above conclusions can also be made about a nonconvex quadrilateral (the reader should check this with tracing paper). Pictures of the two tessellations generated in this manner are shown in Fig. 3.12. These observations lead to the following generalization: *Any quadrilateral will tessellate the plane.*

The tessellation kit produced by D. Daigger (available from Educational Teaching Aids) and the accompanying work cards provide useful materials for exploring tessellations. Students should be encouraged to think about the sides and angles of the various polygons as they investigate tessellation properties.

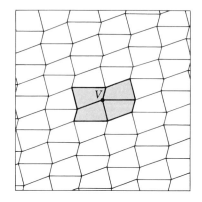 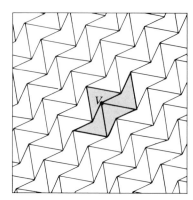

Figure 3.12

In summary, we have learned that (a) equilateral triangles, squares, and regular hexagons are the only regular polygons which tessellate the plane, (b) all triangles tessellate the plane, and (c) all quadrilaterals tessellate the plane.

Since all triangles and quadrilaterals tessellate the plane, it is natural to ask the question, What types of pentagons and hexagons will tessellate the plane?

While a regular pentagon will not tessellate, it is interesting to note that there is a pentagon (see region A in Fig. 3.13) with all sides congruent (but with different-size angles) that will tessellate the plane. A portion of this tessellation is shown in Fig. 3.13.

If four of these pentagonal regions are considered together (see region B), an interesting hexagonal shape results which will tessellate the plane.

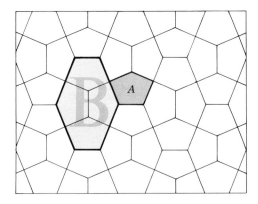

Figure 3.13

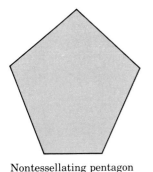

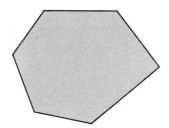

Nontessellating pentagon

Nontessellating hexagon

Figure 3.14

Thus we have exhibited a nonregular pentagon and a nonregular hexagon which *will* tessellate the plane. The pentagon and the hexagon shown above, however, are examples of these types of figures which *will not* tessellate the plane. *Scientific American* (July 1975) presents Richard B. Kershner's descriptions of all possible tessellating pentagons (8 types) and all possible tessellating hexagons (3 types).

It can be proven that no *convex* polygon with more than six sides will tile the plane. However, other irregular polygons will tessellate the plane, and some interesting methods can be used to construct these tessellations. Figure 3.14 illustrates how a tessellation of irregular shapes can be constructed from two tessellations, one superimposed upon the other. Figure 3.15 shows other interesting types of tessellations.

Teachers of mathematics are continually involved in attitude development. Work with tessellations provides an avenue for creative interest in geometry and helps students see mathematics as an enjoyable activity. Also, work with tessellations provides an avenue for integrating geometry and art.

Figure 3.15

It is possible to tessellate the plane with polygons shaped like certain letters of the alphabet.

A hexaminoe is a figure made of six squares. Certain hexaminoes will tessellate the plane.

Certain odd-shaped figures that one wouldn't expect to tessellate the plane provide interesting tessellations.

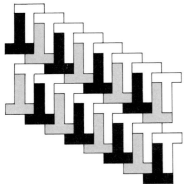

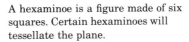

EXERCISE SET 3.1

1. a) On square dot paper draw a tessellation using this triangle as the basic figure.
   b) Draw in red the vertex figures for the tessellation you drew in part (a).
   c) Are these vertex figures on your tessellation of one single shape? If not, redraw your tessellation so that there is only one shaped vertex figure.

2. On square dot paper draw a tessellation using this quadrilateral.
   Make sure that your tessellation possesses only one type of vertex figure.

3. Using a trapezoid with the following shape, construct:
   a) a tessellation with this trapezoid so that each vertex figure is a kite, and
   b) a second tessellation in which the vertex figures are not kites.

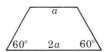

4. Using dot paper draw, if possible, tessellations of each of the pentominoes given here. (The definition of a pentominoe is given in Chapter 2, p. 46.)

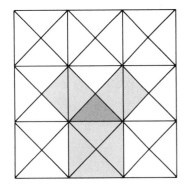

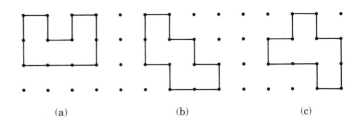

(a)   (b)   (c)

5. The shaded portion of the tessellation at the left suggests what famous theorem from plane geometry?

6. Each figure on the tessellation of these triangles suggests at least one theorem from plane geometry. State at least one theorem for each part.

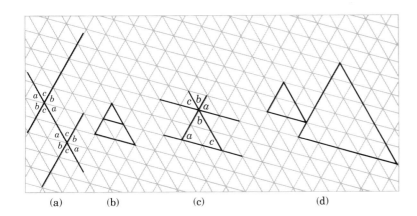

(a)   (b)   (c)   (d)

7. An equilateral triangle is called a "reptile" (an abbreviation for "repeating tile") because four equilateral triangles can be arranged to form a larger equilateral triangle. Which of these figures are "reptiles"? Use appropriate dot paper and show the larger figure in each case.

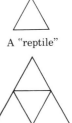

A "reptile"

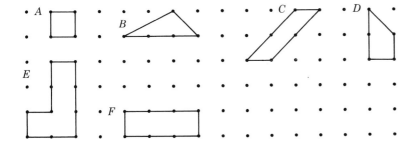

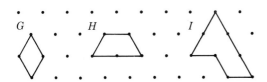

8. Figure 3.15 shows a tessellation using the letter  as the basic figure. Which of these letters of the alphabet will tessellate the plane?

Show a portion of the tesselation(s).

9. The *dual* of a tessellation is a new tessellation obtained by connecting the centers (centroids) of polygons that share a common side. The dual (shown in gray) of the tessellation of regular hexagons is the tessellation 3^6. Describe the dual of the regular tessellation of (a) squares, (b) equilateral triangles.

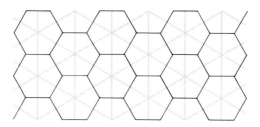

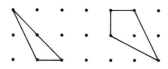

10. Given the triangle and quadrilateral (left) we ask:

   a) What is the maximum number of these triangles which can be drawn without overlap on square dot paper with seven dots on a side?

   b) What is the maximum number of these quadrilaterals which can be drawn without overlap on square dot paper with seven dots on a side.

*11. There are five tetrominoes (pentominoe-like figures made up of four squares).

   Draw pictures of these tetrominoes and decide which of these will tessellate the plane.

*12. There are twelve pentominoes.

    a) Draw pictures of these pentominoes and decide which of them will tessellate the plane.

    b) Use one copy of each of the pentominoes to "tile" a rectangle region with length 10 and width 6 (the unit of length is the length of one side of one of the squares that form the pentominoes).

*13. There are thirty-five hexominoes (pentominoe-like figures made up of six squares). Find at least three hexominoes that will tessellate the plane.

*14. There are four pentiamonds. (A pentiamond is a figure made up of five connected equilateral triangles in such a way that each triangle always has at least one side in common with an adjacent triangle.) Here is an example.

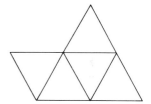

Draw a picture of each pentiamond and decide which of these pentiamonds will tessellate the plane.

*15. There are twelve hexiamonds (figures made from six equilateral triangles in the manner described in Exercise 14). Find at least two hexiamonds that will tessellate the plane.

Exercises 16 through 21 describe a collection of pentagons and hexagons which tessellate.

16. a) Use tracing paper and trace four copies of this tessellation. Draw around two adjacent quadrilaterals to form a different-shaped hexagon on each copy.

    b) What does the experience in part (a) suggest about finding hexagons which tessellate the plane.

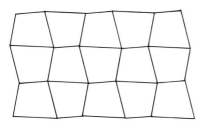

*17. Consider a parallelogram $ABCD$ and let x and y be two interior points. Construct points x_1 and x_2 so that A is the midpoint of xx_1 and B is the midpoint of xx_2. Construct y_1 and y_2 so that D is the midpoint of yy_1 and C is the midpoint of yy_2. The hexagon $x_1xx_2y_2yy_1$ will always tessellate the plane. Use tracing paper and accurate drawings to convince yourself that this is true.

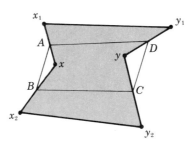

*18. Consider any quadrilateral $ABCD$. Find the midpoint M of any side (i.e., AD) of the quadrilateral and draw any segment PQ such that P is inside $ABCD$ and M is the midpoint of PQ. Draw segments PD and AQ. The hexagon $ABCDPQ$ will always tessellate the plane. Try this with a quadrilateral of your choice and use tracing paper to draw a portion of the tessellation to convince yourself that this is true.

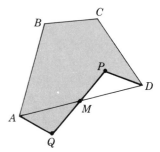

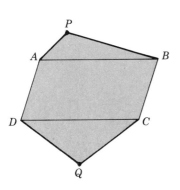

*19. Consider a parallelogram $ABCD$ and choose points P and Q outside $ABCD$ such that $\angle PAB$, $\angle PBA$, $\angle QDC$, and $\angle QCD$ are all acute angles. Draw PA, PB, QD, and QC. The hexagon $PBCQDA$ will always tessellate the plane.

Try this with a parallelogram and points P and Q of your choice. Use tracing paper to draw a portion of the tessellation.

*20. Draw any pair of intersecting lines l and m and a pair of intersecting lines l' and m' which are parallel to l and m. Choose any points P and Q on l and l' and points R and S on m and m'. Connect points P and Q and R and S. The hexagon $PTRSVQ$ will always tessellate the plane. Try this with lines of your own. Use tracing paper to draw a portion of the tessellation.

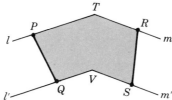

*21. Draw any triangle ABC. Find the midpoint M of any side (i.e., AC) of the triangle and draw any segment PQ such that P is inside the Δ and M is the midpoint of PQ. Draw segments PC and QA. Pentagon $ABCPQ$ will always tessellate the plane. Try this with a triangle of your choice. Using tracing paper to draw a portion of the tessellation.

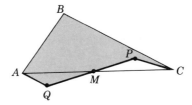

TESSELLATIONS WITH REGULAR POLYGONS

Combinations of several polygon shapes often yield tessellations with interesting and beautiful patterns, patterns with artistic appeal (Fig. 3.16). However, in order to concentrate on the mathematics of tessellations and not on the art in tessellations, we restrict our discussion in the remainder of this chapter to tessellations formed by combinations of *regular* polygons. Two examples are shown in Fig. 3.17.

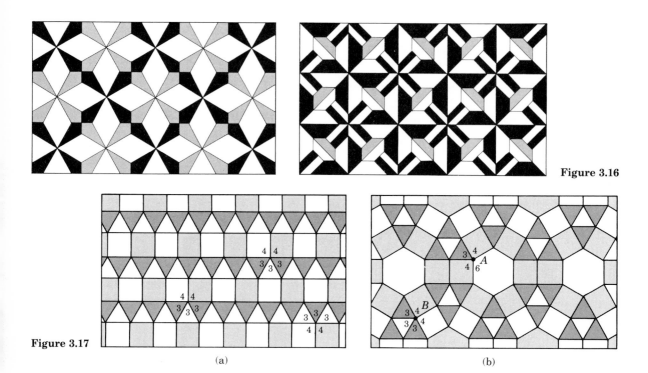

Figure 3.16

Figure 3.17

(a)   (b)

One of the purposes for studying geometry is to discover order and pattern in the universe. Students should be encouraged to look for examples of tessellations in their environment. A few of the situations in which they occur are in floor tilings, in wallpaper designs, in brick walls, in architectural patterns, and in various aspects of nature.

There are many tessellations which are formed by a combination of several regular polygons, and these tessellations can be classified according to certain properties. Note that the tessellation in Fig. 3.17(a) possesses only one type of vertex figure while the tessellation in Fig. 3.17(b) possesses two types of vertex figures.

The tessellation in Fig. 3.17(a) is an example of a type of tessellation called a semiregular tessellation. A tessellation is a **semiregular tessellation** if it is composed of regular polygons of two or more types in such a way that all vertex figures are identical; that is, the arrangement of polygons at each vertex is the same. Why is the tessellation in Fig. 3.17(b) not semiregular?

In the preceding section of this chapter we have seen that there are only three regular tessellations. In this section we answer the question, *How many semiregular tessellations are there?* As a first step in answering this question, we search for

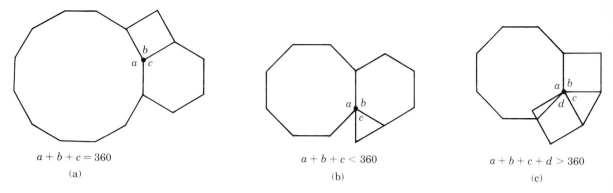

$a + b + c = 360$

(a)

$a + b + c < 360$

(b)

$a + b + c + d > 360$

(c)

Figure 3.18

combinations of regular polygons which completely surround a vertex with no overlapping. For example, we see that a regular 12-gon, 6-gon, and 4-gon surround a vertex with no overlapping (Fig. 3.18a) whereas a regular 8-gon, 6-gon, and 3-gon leave a gap (Fig. 3.18b) and a regular 8-gon, 4-gon, 3-gon, and 4-gon around a vertex overlap (Fig. 3.18c). Once we find some combinations of regular polygons which completely surround a vertex, we shall determine which of these vertex arrangements can be extended to tessellate the plane.

In Investigation 3.2 we shall find some combinations of regular polygons which surround a vertex with no overlapping.

Investigation 3.2

Cut out several copies of an equilateral triangle, a square, a regular hexagon, a regular octagon, and a regular dodecagon with sides all the same length.

In how many different ways can you fit *three* of these polygons around a point? (A particular type of polygon may be repeated in the arrangement.)

Answer this question for *four, five, six,* and *more than six* polygons surrounding a point.

Use tracing paper on the figures above to trace a picture to record each way you found. Devise a notation to describe each arrangement.

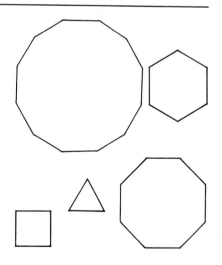

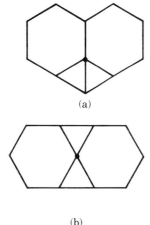

(a)

(b)

Figure 3.19

The following question arises as a result of Investigation 3.2: Should different orderings of the same set of polygons be considered different ways of surrounding a vertex? For example, are the arrangements in Fig. 3.19(a) and (b) the same or different? It is natural to consider them different, since one figure has one line of symmetry and the other has two lines of symmetry. In summary, when counting different orderings as different arrangements, the reader should have found the following while completing Investigation 3.2:

Using only an equilateral triangle, a square, a regular hexagon, a regular octagon, and a regular dodecagon, there are

a) four ways of surrounding a vertex with three of these polygons,

b) seven ways of surrounding a vertex with four of these polygons,

c) three ways of surrounding a vertex with five of these polygons,

d) one way of surrounding a vertex with six of these polygons, and

e) zero ways of surrounding a vertex with more than six of these polygons.

Later in this chapter we will verify that a vertex can also be surrounded by

a) two 5-gons and a 10-gon;

b) a square, a 5-gon, and a 20-gon;

c) an equilateral triangle, a 7-gon, and a 42-gon;

d) an equilateral triangle, an octagon, and a 24-gon;

e) an equilateral triangle, a 9-gon, and an 18-gon; and

f) an equilateral triangle, a 10-gon, and a 15-gon.

(The reader may verify these by calculating the number of degrees in the vertex angle of each polygon and checking to see if the angles around a vertex sum to 360.)

We shall also show that *the twenty-one arrangements described above are the only possible arrangements of regular polygons around a vertex.*

Since three of these arrangements produce the familiar regular tessellations (4,4,4,4; 3,3,3,3,3,3; and 6,6,6), we are left with *eighteen* possibilities to consider. A crucial question immediately arises: *Can every one of these eighteen arrangements of regular polygons around a single vertex be extended to form a semiregular tessellation?* The answer is "no." For example, consider an arrangement of regular polygons around vertex V which we shall denote by 6,6,3,3. If we attempt to extend this configuration so that the same arrangement will occur around vertex V_1, we now find that it will be impossible to effect this same arrangement around V_2. Thus the arrangement 6,6,3,3 cannot be extended to produce a semiregular tessellation.

Which of the vertex arrangements found in Investigation 3.2 cannot be extended to form a semiregular tessellation? This question motivates our next investigation.

Investigation 3.3

If we exclude the regular tessellations, there are four of the arrangements you found in Investigation 3.2 which cannot be extended to form a semiregular tessellation.

Can you find them and explain why they cannot be extended? (Tracing paper and the figures in Investigation 3.2 may be helpful.)

After eliminating in Investigation 3.3 four of the arrangements found in Investigation 3.2, the remaining arrangements of 3-gons, 4-gons, 6-gons, 8-gons, and 12-gons produce eight semiregular tessellations (Fig. 3.20). These eight are the only semiregular tessellations that can be formed by using the regular polygons from Investigation 3.2.

There are questions concerning semiregular polygons which remain unanswered. For example, can any of the other arrangements of regular polygons about a point, which include pentagons or other n-gons, be extended to form semiregular tessellations? How many semiregular tessellations are there in all? We explore these questions in the next section.

Ditto copies of the eight semiregular tessellations can be made and given to students to color in an interesting way. Students at different ages and with varying abilities in mathematics enjoy the artistic aspects of this activity, and the ability to recognize geometric patterns is enhanced. Older students can try to create tessellations with different reflectional symmetry properties when color is considered.

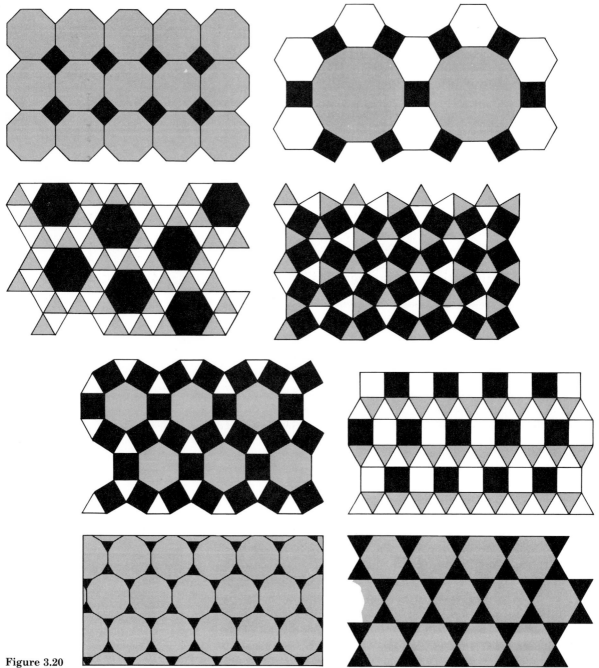

Figure 3.20

ANALYSIS OF SEMIREGULAR TESSELLATIONS*

In this section we proceed with a more careful analysis of semiregular tessellations so that at the end of the section we can declare that all semiregular tessellations have been found.

We have seen that a vertex angle of a regular n-gon measures $180\left(1 - \dfrac{2}{n}\right)$ degrees. If there are three regular polygons, an n_1-gon, an n_2-gon, and an n_3-gon which completely surrounds a vertex with no overlapping, (Fig. 3.21), then the three numbers (n_1, n_2, n_3) must satisfy the equation.

i) $$180\left(1 - \frac{2}{n_1}\right) + 180\left(1 - \frac{2}{n_2}\right) + 180\left(1 - \frac{2}{n_3}\right) = 360$$

Conversely, any set of whole-number solutions to equation (i) represents three regular polygons which surround a vertex with no overlapping. In searching for all whole number solutions to equation (i), it is helpful to use the equivalent equation (ii) which we derive.

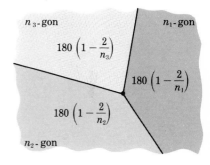

Figure 3.21

i) $$180\left(1 - \frac{2}{n_1}\right) + 180\left(1 - \frac{2}{n_2}\right) + 180\left(1 - \frac{2}{n_3}\right) = 360$$

$$\Leftrightarrow \left(1 - \frac{2}{n_1}\right) + \left(1 - \frac{2}{n_2}\right) + \left(1 - \frac{2}{n_3}\right) = 2$$

$$\Leftrightarrow 3 - \left(\frac{2}{n_1} + \frac{2}{n_2} + \frac{2}{n_3}\right) = 2$$

$$\Leftrightarrow \frac{2}{n_1} + \frac{2}{n_2} + \frac{2}{n_3} = 1$$

ii) $$\Leftrightarrow \frac{1}{n_1} + \frac{1}{n_2} + \frac{1}{n_3} = \frac{1}{2}$$

Since there can also be four, five, or six regular polygons surrounding a vertex with no overlapping, we also need to consider solutions to the three additional equations recorded in Table 3.2 which are similar to equation (i).

---

* An asterisk beside a section title indicates that the section is less basic to an understanding of the material that follows, or is more difficult than usual and may be omitted in a minimal course.

We seek all whole number solutions to the equations in the above table. Each solution describes a set of regular polygons which completely surround a vertex with no overlapping, and hence each solution represents a potential vertex arrangement in a semi-regular tessellation. There is a total of seventeen solutions to the four equations in the right column of Table 3.2, all of which are listed in Table 3.3. By considering the vertex arrangements represented by each solution we can find all semiregular tessellations. Note that the solutions given in lines 12, 13, 14, and 16 can be considered in different orders to produce different vertex arrangements.

A careful study of the solutions shown in Table 3.3 and the pictures of the eight semiregular tessellations found so far (Fig. 3.20) provides the following information.

1. The solutions in lines 1, 11, and 17 represent the vertex arrangements of three regular tessellations.

Table 3.2

Number of polygons surrounding a vertex	Resulting equation	Equivalent equation
3	$180\left(1-\dfrac{2}{n_1}\right) + 180\left(1-\dfrac{2}{n_2}\right) + 180\left(1-\dfrac{2}{n_3}\right) = 360$	$\dfrac{1}{n_1} + \dfrac{1}{n_2} + \dfrac{1}{n_3} = \dfrac{1}{2}$
4	$180\left(1-\dfrac{2}{n_1}\right) + 180\left(1-\dfrac{2}{n_2}\right) + 180\left(1-\dfrac{2}{n_3}\right)$ $+ 180\left(1-\dfrac{2}{n_4}\right) = 360$	$\dfrac{1}{n_1} + \dfrac{1}{n_2} + \dfrac{1}{n_3} + \dfrac{1}{n_4} = 1$
5	$180\left(1-\dfrac{2}{n_1}\right) + 180\left(1-\dfrac{2}{n_2}\right) + 180\left(1-\dfrac{2}{n_3}\right)$ $+ 180\left(1-\dfrac{2}{n_4}\right) + 180\left(1-\dfrac{2}{n_5}\right) = 360$	$\dfrac{1}{n_1} + \dfrac{1}{n_2} + \dfrac{1}{n_3} + \dfrac{1}{n_4} + \dfrac{1}{n_5} = \dfrac{3}{2}$
6	$180\left(1-\dfrac{2}{n_1}\right) + 180\left(1-\dfrac{2}{n_2}\right) + 180\left(1-\dfrac{2}{n_3}\right)$ $+ 180\left(1-\dfrac{2}{n_4}\right) + 180\left(1-\dfrac{2}{n_5}\right) + 180\left(1-\dfrac{2}{n_6}\right) = 360$	$\dfrac{1}{n_1} + \dfrac{1}{n_2} + \dfrac{1}{n_3} + \dfrac{1}{n_4} + \dfrac{1}{n_5} + \dfrac{1}{n_6} = 2$

Table 3.3

Equations	Solutions						Line number
	n_1	n_2	n_3	n_4	n_5	n_6	
	6	6	6				1
	5	5	10				2
	4	5	20				3
	4	**6**	**12**				4
$\dfrac{1}{n_1} + \dfrac{1}{n_2} + \dfrac{1}{n_3} = \dfrac{1}{2}$	**4**	**8**	**8**				5
	3	7	42				6
	3	8	24				7
	3	9	18				8
	3	10	15				9
	3	**12**	**12**				10
	4	4	4	4			11
$\dfrac{1}{n_1} + \dfrac{1}{n_2} + \dfrac{1}{n_3} + \dfrac{1}{n_4} = 1$	3	3	4	12			12
	3	**3**	**6**	**6**			13
	3	**4**	**4**	**6**			14
$\dfrac{1}{n_1} + \dfrac{1}{n_2} + \dfrac{1}{n_3} + \dfrac{1}{n_4} + \dfrac{1}{n_5} = \dfrac{3}{2}$	**3**	**3**	**3**	**3**	**6**		15
	3	**3**	**3**	**4**	**4**		16
$\dfrac{1}{n_1} + \dfrac{1}{n_2} + \dfrac{1}{n_3} + \dfrac{1}{n_4} + \dfrac{1}{n_5} + \dfrac{1}{n_6} = 2$	3	3	3	3	3	3	17

2. The solutions in lines 4, 5, 10, 13, 14, 15, and 16 provide the vertex arrangements that can be extended to form the semiregular tessellations in Fig. 3.20. The eight particular arrangements that form the tessellations are given in bold face type below.

Equation solutions	Possible vertex arrangements	
line 4 – 4,6,12	**4,6,12**	
line 5 – 4,8,8	**4,8,8**	
line 10 – 3,12,12	**3,12,12**	
line 13 – 3,3,6,6	**3,6,3,6**	3,3,6,6
line 14 – 3,4,4,6	**3,4,6,4**	3,4,4,6
line 15 – 3,3,3,3,6	**3,3,3,3,6**	
line 16 – 3,3,3,4,4	**3,3,4,3,4**	3,3,3,4,4

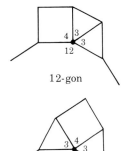

12-gon

12-gon

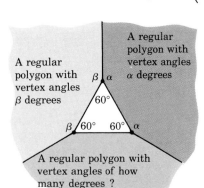

A regular
polygon with
vertex angles
α degrees

A regular
polygon with
vertex angles
β degrees

A regular polygon with
vertex angles of how
many degrees ?

Figure 3.22

3. The solution in line 12, which is 3,3,4,12, provides no vertex arrangement that can be extended to form a semi-regular tessellation. (Recall your work in Investigation 3.3.)

   That is, neither 3,3,4,12 nor 3,4,3,12 describe a semiregular tessellation.

4. It remains to consider the solutions in lines 2, 3, 6, 7, 8, 9 of Table 3.3.

First, let us consider lines 6 through 9. Each of these solutions suggests a vertex arrangement in which an equilateral triangle and two other regular polygons surround a vertex. Figure 3.22 suggests that this type of arrangement cannot be extended to form a semiregular tessellation unless the other two regular polygons are the same. Note that if the "lower" polygon has vertex angles α, the arrangements at each vertex won't be the same. Similarly, if it has vertex angles β, the arrangements at each vertex won't be the same. Thus it is impossible to extend this arrangement to form a semiregular tessellation unless $\alpha = \beta$. Because of this, none of

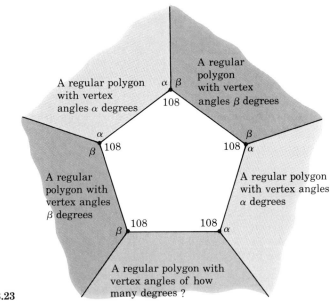

A regular polygon
with vertex
angles α degrees

A regular
polygon
with vertex
angles β degrees

A regular
polygon with
vertex angles
β degrees

A regular polygon
with vertex angles
α degrees

A regular polygon with
vertex angles of how
many degrees ?

Figure 3.23

the solutions in lines 6 through 9 represent semiregular tessellations, since no two of the polygons are the same.

In a similar manner, Fig. 3.23 suggests that an arrangement with a regular pentagon and two other regular polygons about each vertex cannot be extended to form a semiregular tessellation unless the other two regular polygons are the same. Thus it is impossible to extend this arrangement to form a semiregular tessellation unless $\alpha = \beta$. Since the two polygons other than the pentagon in lines 2 and 3 are not the same, these solutions do not represent semiregular tessellations.

We conclude that the eight semiregular tessellations in Fig. 3.20 are the *only* semiregular tessellations.

EXERCISE SET 3.2

1. The notation 3,6,3,6 is often used to represent the semiregular tessellation with vertex arrangement of an equilateral triangle, regular hexagon, equilateral triangle, and regular hexagon in that order. Represent each of the remaining seven semiregular tessellations using a similar notation.

2. We have seen in the figure on page 95 that the vertex arrangement shaded below cannot be extended to a semiregular tessellation. Conversely, the drawing below

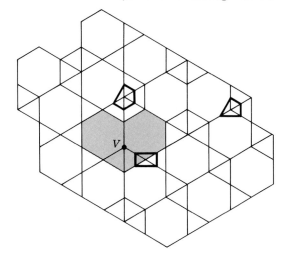

illustrates that the vertex arrangement can be extended to a nonsemiregular tessellation. Select another one of the vertex arrangements found in Investigation 3.3 which cannot be extended to a semiregular tessellation and determine whether or not it can be extended to a non-semiregular tessellation.

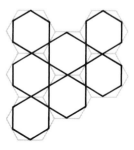

Vertex figure tessellation

3. The figure (left) illustrates that two new tessellations can be constructed from any tessellation. The *vertex figure* tessellation is constructed by drawing all the vertex figures of the original tessellation, and the *dual* tessellation is constructed by connecting the centers (centroids) of neighboring polygons.

a) Choose one of the semiregular tessellations of Fig. 3.20 and trace it on tracing paper. Draw (in a second color) the dual tessellation. Is this dual a regular tessellation, a semiregular tessellation, or neither?

b) Choose a semiregular tessellation different from the one chosen in Exercise 3(a). Trace this tessellation and draw in all the vertex figures (using a second color). Is the new tessellation, made up of only the vertex figures, a semiregular tessellation?

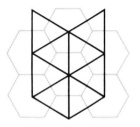

Dual tessellation

THREE-MIRROR KALEIDOSCOPE

Students of all ages find the three mirror kaleidoscope fascinating. The creation of patterns for the kaleidoscope provides an experimental situation which encourages a creative approach to design. This activity also provides a situation rich in opportunities for a deeper look into properties of polygons, including angles, sides, diagonals, and lines of symmetry. The Pattern Blocks and Mirrors kit (by Selective Educational Equipment, Inc.) is excellent for experimenting with tessellations and kaleidoscopes.

It is both exciting and instructive to generate tessellations using mirrors. Fasten three mirrors together to form the vertical faces of an equilateral triangular prism. Place construction paper or colored acetate patterns in the base of the kaleidoscope. The viewer (Fig. 3.24) beholds an interesting and sometimes beautiful tessellation as he peers over the edge. Try it!

Figure 3.24

The tessellation in Fig. 3.25 was generated by the indicated pattern being placed in the base of a kaleidoscope.

Investigation 3.4 provides an experience which should help build the reader's intuition about symmetry while generating some interesting tessellations.

Figure 3.26 may provide a hint for parts B and C of this investigation.

Figure 3.25

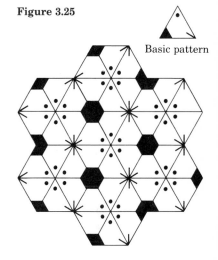

Basic pattern

Figure 3.26

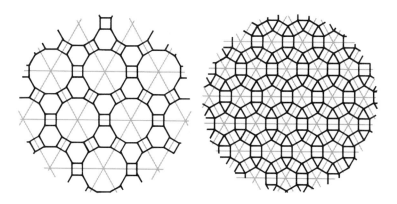

Investigation 3.4

A. Each of the patterns below when placed in the base of a three-mirror kaleidoscope generates a tessellation.

For each pattern, guess what the tessellation will look like.

Check your guess with a kaleidoscope.

B. Can the regular tessellations be generated by placing patterns in a 3-mirror kaleidoscope? If so, show the patterns.

C. How many of the semiregular tessellations can be generated by placing patterns in a three-mirror kaleidoscope. Show the patterns.

TESSELLATIONS OF CURVED PATTERNS

It is possible to create elaborate and interesting tessellations formed by patterns of curved figures. In Fig. 3.27 we see some examples of this created by the artist Maurits Escher.

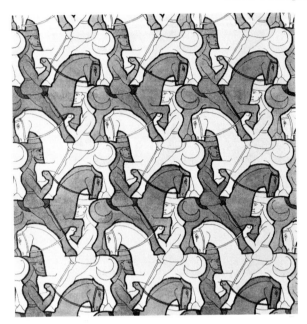 

Drawings, M. C. Escher

Figure 3.27

Activities such as these can sometimes catch the interest of students who haven't been very successful in other aspects of mathematics. The book *Tessellations*, Cambridge University Press, by Josephine Mold, suggests further ideas for student activities. Also the April, 1974 issue of *The Mathematics Teacher* will provide further suggestions for students.

Many of these tessellations of curved figures can be formed by modifying a tessellation of polygons. One approach involves replacing one or more edges of each polygon with a curved line. In Fig. 3.28 we see how the regular tessellation of equilateral triangles can be transformed into a tessellation figure.

A second approach involves providing the interior of each polygon with some pattern or design. For example, a square designed as in Fig. 3.29 generates the tessellation of Fig. 3.30. In Fig. 3.31 we see how a tessellation of quadrilaterals can be converted into two shoals of fish swimming in opposite directions.

The interested reader might wish to create some of his own tessellations by employing these techniques.

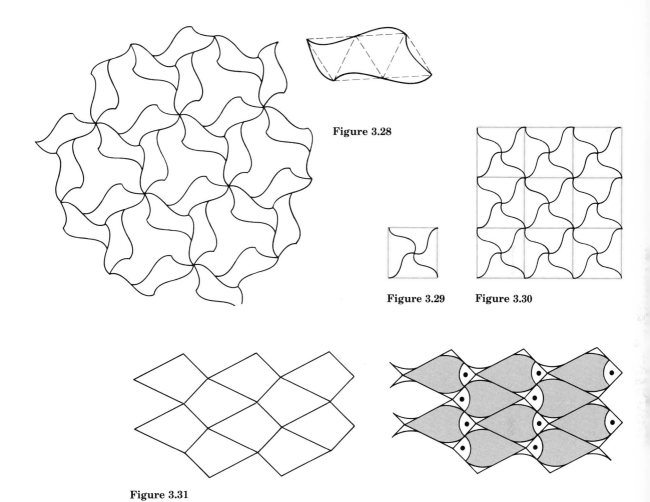

Figure 3.28

Figure 3.29   Figure 3.30

Figure 3.31

EXERCISE SET 3.3

Using one of the methods illustrated in Figures 3.28 through
3.31 draw a tessellation of curved figures.

BIBLIOGRAPHIC REFERENCES

[4, 19–22], [5, 43–54], [6, 131–143, 157–160], [7], [16], [26, 50–65],
[32], [34, 46–49, 75, 146–161], [35], [40, 248–252], [43, 222–223],
[47], [59], [63], [66], [70], [71], [85], [86], [87], [88], [92], [95],
[97], [100], [102], [109].

PEDAGOGICAL ACTIVITIES FOR THE TEACHER

1. Procure some laboratory materials such as "tessellations," produced by Daigger (Educational Teaching Aids), and explore the activity cards for these materials. Then use the materials with a student at an appropriate level. Summarize the student's reaction to the materials. How would you improve this experience for students?

2. Use the Pattern Blocks (produced by McGraw-Hill) and write five questions which would encourage a student to use these materials to explore some aspects of tessellations. Try these questions with a student at an appropriate level. Then revise the questions for future use.

3. Play games such as Psyche Paths, available from Cuisenaire Company of America, Inc. Develop criteria for evaluation and evaluate these games according to that criteria. Your main goal should be to decide the value of these games as a learning experience for students.

4. Devise an investigation card that has the potential to get upper grade or junior high school students involved in a search for all the semiregular tessellations.

5. Discuss tessellations with someone from the art department in your school and list some ideas for integrating these ideas of mathematics with an art project.

6. Devise a way to capture a student's interest and get him involved in some work with tessellations. The work might include coloring tessellations in interesting ways, making curved-line tessellations, searching for semiregular tessellations, etc. Your means of motivating the student might range from a photography project to an art contest.

7. Analyze a representative elementary school or upper-level mathematics text series in the following way: (a) check each grade level and see if any activities which involve tessellations are included; (b) if the text series does not include these activities, list the page number of specific places in the series in which you feel such activities might be included. Be specific about what activities you would include.

8. Make a set of worksheets which include some or all of the semiregular tessellations and present them to primary school children. Ask the children to color them in interesting ways. Display the results for your fellow teachers to observe and analyze.

9. Write a sequence of discovery questions which would lead a student to use tessellations to discover the theorems suggested in Exercise 6 on page 87.

10. Read an article such as "Kaleidoscope Geometry" in February, 1970 *Arithmetic Teacher,* and devise a specific lesson plan including behavioral objectives which you can use in a fifth or sixth grade classroom and which involves activities using a kaleidoscope.

4

Polyhedra and Tessellations of Space

INTRODUCTION

Plato (427–348 B.C.), who required all his pupils to obtain a thorough knowledge of mathematics before being initiated into his philosophy, was asked "What does God do?" He is said to have responded, "God eternally geometrizes". It is in the natural context of our three-dimensional environment that we consider the impact of Plato's statement. Whether or not we agree, we must surely sense profound wonderment when we consider the myriad of exciting shapes that occur in the microscopic world of matter; in the natural world of rocks, plants, and animals; in the created world of the architect and artist; and in the vast world of outer space which we can barely see. If we but pay attention to the crystals under a microscope, to the skeletons of tiny sea creatures, to shells, sunflowers, and honeybees, and to the human body and the cathedrals man has built, we might find that a course in geometry would be an afterthought.

Yet we often do not have the inclination nor the opportunity to be so observant. In fact, many very simple aspects of geometry in familiar everyday situations often escape us. To begin our study of three-dimensional space, we consider one of those situations. The reader who has observed a brick wall being built or a child playing with blocks may not have thought of this simple situation as a source of some rather interesting geometrical questions. The following investigation will provide an opportunity to explore the spatial relationships involved in wall building.

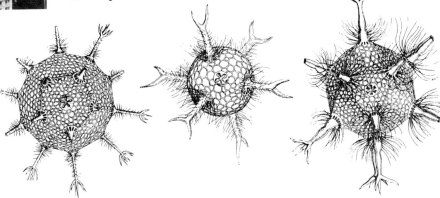

The skeletons of tiny sea creatures called radiolarians form elaborate geometric shapes.

Investigation 4.1

A. For each condition given below, build a wall (theoretically of unlimited length and height) by repeating a basic pattern. (Use either children's building blocks or rectangular cardboard cutouts which simulate blocks.)

Build a wall:

1. with layers which can shift (that is, the wall can be sliced with a horizontal plane without cutting any bricks);

2. that can be sliced with both horizontal and vertical planes without cutting any bricks;

3. that cannot be sliced with a plane without cutting bricks (except, of course, the plane of the table on which you are working);

4. in such a way that each brick in the center of the wall touches five neighboring bricks (Two bricks are called "neighboring" only if they touch along a flat surface. For example, the brick here has four neighbors.);

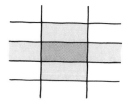

5. So that each brick in the center of the wall touches six neighboring bricks.

B. Sketch a picture of each of your walls.

Every school classroom should have a set of geometry solids made from wood, styrofoam, plexiglass, or cardboard. These forms provide the very small child with a rich environment in which spatial perceptions of polyhedra are developed by comparing and contrasting the models, using touch and sight.

The older child uses the models to explore quantitative questions, such as finding volumes and counting faces, edges, and vertices.

Extending this investigation by considering walls of at least three bricks thick (with every other layer shifted slightly), we see in the photographs below that it is possible for a brick to have 14 neighbors. This fact is related to a space tessellation which will be described later in this chapter.

The "bricks" you used in Investigation 4.1 exemplify one of the simplest of an important class of solid figures, called **polyhedra**. A **polyhedron** can be thought of as a finite set of polygons joined pairwise along edges to enclose a finite region of space. If this region is a convex region, the polyhedron is called a **convex polyhedron**. (A convex region is defined in a manner similar to that used in defining a convex polygon.)

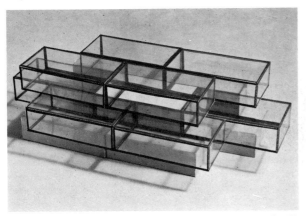

From *Mathematical Snapshots* by H. Steinhaus, 3rd edition. Copyright © 1950, 1960, 1968, 1969 by Oxford University Press, Inc. Reprinted by permission.

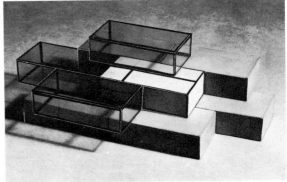

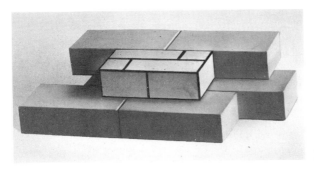

Figure 4.1

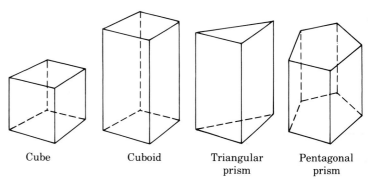

Cube	Cuboid	Triangular prism	Pentagonal prism

Figure 4.2   Right regular prisms

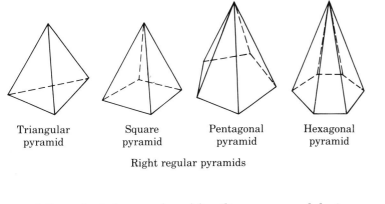

Triangular pyramid	Square pyramid	Pentagonal pyramid	Hexagonal pyramid

Right regular pyramids

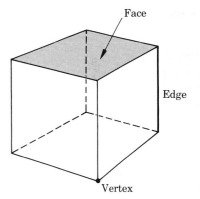

A few selected examples of familiar convex polyhedra are illustrated and named in Fig. 4.2.

The polygons forming a polyhedron are called **faces**. Any common side of two faces is called an **edge**. The points of intersection of edges are called **vertices**.

REGULAR POLYHEDRA

In the book *The Hunting of the Snark* by Lewis Carroll (an alias for C. L. Dodgson, the mathematician-writer who wrote *Alice in Wonderland*), the following appears:

> You boil it in sawdust: you salt it in glue:
> You condense it with locusts and tape:
> Still keeping one principal object in view—
> To preserve its symmetrical shape.

Although the following investigation does not require such a drastic operation, it does provide an opportunity for some simple model building. It has been said that the best way to learn about polyhedra is to make them; the next best way is to handle them. This investigation will provide an opportunity for both. Of course, in addition to the experience of manipulating physical objects, the reader should be seeking conjectures about any relationships observed.

Investigation 4.2

One class of polyhedra includes those in which all faces are the same size and shape.

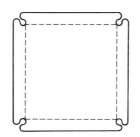

A. Use commercially produced materials or cut from poster-board as many square patterns shaped like this as you need and put them together with rubber bands to form a cube. Can you construct any other solid figure with square faces?

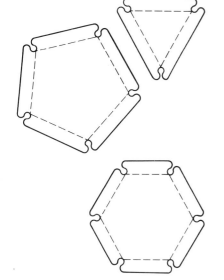

B. Here are some other regular polygon patterns.

   Can you use larger patterns like these and make other polyhedra with each face the same shaped polygon? How many different such polyhedra can you make?

C. Can you make polyhedra from any other regular polygon patterns?

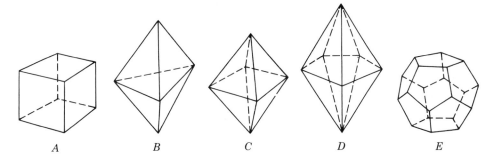

Figure 4.3   *A*   *B*   *C*   *D*   *E*

Some of the polyhedra you may have made in Investigation 4.2 are pictured in Fig. 4.3. In these polyhedra, each face is a regular polygon and all faces are congruent (same size and same shape). Figures *A*, *C*, and *E* have an important characteristic, however, that is not found in Figs. *B* and *D*.

In order to focus on this characteristic, recall that in Chapter 3, we discussed plane tessellations of regular polygons. The tessellation shown in Fig. 4.4, consisting of pentagons with all sides equal, is often mistaken for an example of a tessellation of the plane with regular pentagons. However, to have a regular polygon, all vertex angles must be of equal measure as well as all sides being of equal length. Similarly, in space a *regular polyhedron* is characterized not only by the condition that the faces are congruent regular polygons, but also by the further condition that the edges and vertices are all "alike." This can be described more precisely by saying that all vertex figures must be regular polygons of the same type. (A vertex figure of a polyhedron is a polygon whose vertices are midpoints of the edges emanating from a vertex of the polyhedron.) In the cube below, the vertex figure is an equilateral triangle. With this condition in mind, the reader should decide which polyhedra in Fig. 4.3 would be classified as regular polyhedra.

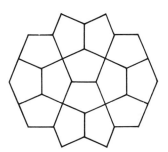

Figure 4.4

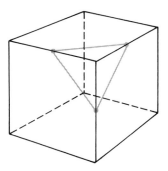

While younger children develop concepts of point, line, and plane by feeling vertices, edges, and faces of polyhedra models, older students enjoy building the models themselves. Books by Marcus Wenniger, entitled *Polyhedral Models for the Classroom* and *Polyhedra Models,* provide some very good activities for students of all ages.

Now that the characteristics of regular polyhedra have been specified, a natural question arises: How many regular polyhedra are there? Perhaps the experiences from Investigation 4.2 will help. First we shall consider regular polyhedra with equilateral triangular faces. We discovered when studying tessellations that six equilateral triangles completely surrounded a vertex in the plane, as shown in Fig. 4.5. So any regular polyhedron can have at most five equilateral triangular faces surrounding a vertex.

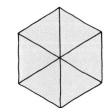

Figure 4.5

Beginning with five equilateral triangles, as shown in Fig. 4.6, and joining points A and B, we produce a three-dimensional "roof," which is a vertex arrangement for a regular polyhedron of twenty faces.

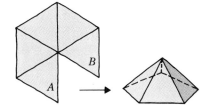

Figure 4.6

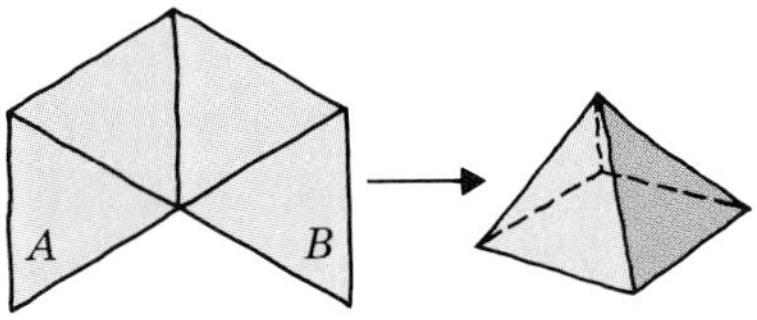

Figure 4.7

Beginning with four equilateral triangles, as shown in Fig. 4.7, and joining corners A and B, we form a three-dimensional "roof" and find that two of these can be placed together to form an eight-faced regular polyhedron.

Beginning with three equilateral triangles, as in Fig. 4.8, and joining corners A and B, we form a three-dimensional "roof" and find that one more equilateral triangle may be placed with this "roof" to form a four-faced regular polyhedron.

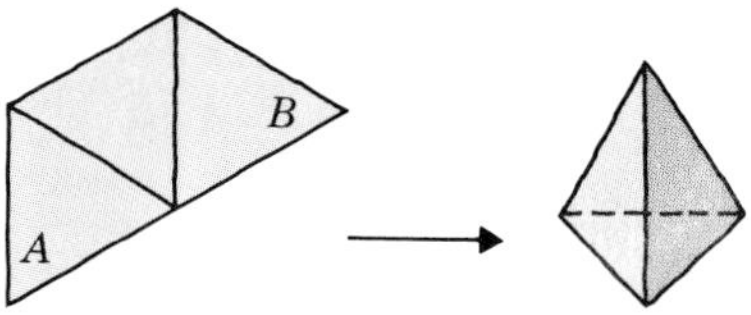

Figure 4.8

Next we consider polyhedra with square faces. Since a vertex in the plane can be surrounded by four squares as in Fig. 4.9, we see that a regular polyhedron with square faces can have at most three squares at each vertex.

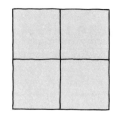

Figure 4.9

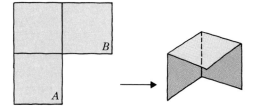

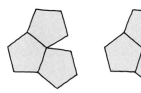

Figure 4.10

Suppose we remove one square from the four in Fig. 4.9 and join corners A and B (Fig. 4.10). We form a three-dimensional "roof" and find that two of these "roofs" may be placed together to form a regular polyhedron with six square faces — the cube. This is the only regular polyhedron with square faces.

The plane cannot be tessellated with regular pentagons because three about a point leave a gap and four about a point overlap, as shown in Fig. 4.11.

Figure 4.11

Figure 4.12

With three pentagons, however, a three-dimensional "roof" can be formed (Fig. 4.12) and it is found that four of these "roofs" will fit together to form a regular polyhedron with twelve regular pentagons as faces.

When we try to place three hexagons about a point, as in Fig. 4.13, we again find that no three-dimensional "roof" is formed, and hence no polyhedra can be made. For regular polygons with more than six sides we find that three or more such polygons placed together at a point always overlap.

Figure 4.13

This intuitive consideration of "roofs", first using equilateral triangles, then squares, regular pentagons, regular hexagons, and larger polygons, strongly suggests that there are exactly five convex regular polyhedra.

A more formal explanation that there can be no more than five convex regular polyhedra is given below.

Suppose there exists a regular convex polyhedron with each face a regular p-gon, as pictured in Fig. 4.14. We have seen in Chapter 2, Fig. 2.19, that each vertex angle has measure $180 ((p-2)/p)°$. If on the convex polyhedron there are q p-gons surrounding each vertex, the sum of the angle

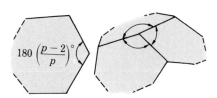

$180 \left(\dfrac{p-2}{p}\right)°$

Figure 4.14

measures around each vertex of the polyhedron would be $q[180((p-2)/p)]$. Note, however, that $q[180((p-2)/p)]$ must be less than 360, for if it is equal to 360, the faces all lie in a plane and we would not have a solid figure. Hence we have:

$$q\left[180\,\frac{(p-2)}{p}\right] < 360$$

$$\Leftrightarrow q\,\frac{(p-2)}{p} < 2$$

$$\Leftrightarrow q(p-2) < 2p$$

$$\Leftrightarrow qp - 2q < 2p$$

$$\Leftrightarrow qp - 2q - 2p < 0$$

$$\Leftrightarrow qp - 2q - 2p + 4 < 4$$

$$\Leftrightarrow (p-2)(q-2) < 4$$

Since a face of the convex regular polyhedron must have more than two sides, and more than two faces are required around each vertex, we see that both p and q must be greater than 2. Hence the following table:

p	q	$p-2$	$q-2$	$(p-2)(q-2)$
3	3	1	1	1
3	4	1	2	2
3	5	1	3	3
4	3	2	1	2
5	3	3	1	3

No other integer values of p and q satisfy the inequality $(p-2)(q-2) < 4$. For if one of p or q is 6 or greater, and the other is at least 3, then one factor in the inequality is 4 or greater and the other is at least 1. In any case, the product $(p-2)(q-2)$ is not less than 4.

Hence we have only five possible convex regular polyhedra: three with equilateral triangular faces—the **tetrahedron, octahedron,** and **icosahedron;** one with square faces—the **cube;** and one with pentagonal faces—the dodecahedron. These polyhedra do exist and are shown with patterns for making them (called **nets**), in Fig. 4.15.

Primary children enjoy an activity in which styrofoam regular polyhedra are placed in a sack and they are asked to reach in the sack and describe the polyhedra simply by feel. It is interesting to see what characteristics of the polyhedra children use in making their description. This activity encourages classification, which is extremely important for young children.

Students at the middle grades and above might be stimulated to look at figures and construct nets rather than vice versa. Also, spatial perception might be tested by presenting nets such as these below, with 6 squares,

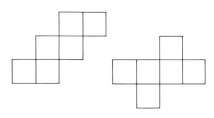

and asking students to decide from the picture whether or not they are nets for a particular polyhedron—in this case, a cube.

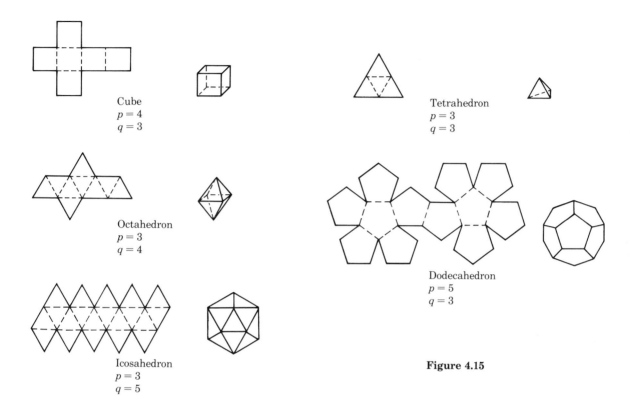

Cube
$p = 4$
$q = 3$

Tetrahedron
$p = 3$
$q = 3$

Octahedron
$p = 3$
$q = 4$

Dodecahedron
$p = 5$
$q = 3$

Icosahedron
$p = 3$
$q = 5$

Figure 4.15

It is interesting to note that nature exhibits the regular convex polyhedra in a variety of exciting ways. Cubes are realized in the shape of salt crystals, while a crystal of pyrite bears a close resemblance to the regular dodecahedron. Crystalline structures as simple as that of carbon are tetrahedral and octahedral in form.

Walter Dawn

Lester Bergman and Associates

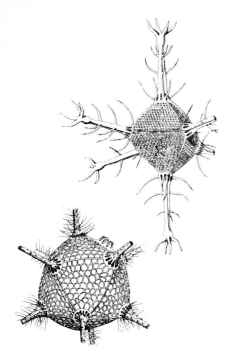

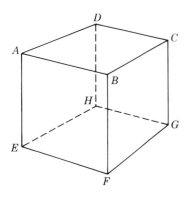

Skeletons of tiny sea creatures, made of silica and measuring only a fraction of a millimeter in diameter, have the form of the octahedron, icosahedron, and the dodecahedron. These creatures, called radiolarians, provide amazingly accurate models of these regular polyhedra. Models of the regular polyhedra occur in modified form in other crystals. For example, the crystals of argentite, the fluorite crystal, resemble octahedra in which vertices have been removed in certain ways.

It is not known who first discovered the regular polyhedra. Perhaps nature was observed and, as is so often the case, the mathematical abstractions followed. It is these abstractions of nature's polyhedra, first investigated carefully by the Greek Theaetetus in 398 B.C., that we consider more carefully in exercises which follow.

EXERCISE SET 4.1

1. An old puzzle asks us to make four triangles using six match sticks. How is a tetrahedron related to this problem?

2. Consider the cube with vertices *ABCDEFGH* shown in the figure:

   a) Two edges of the cube are called opposite if they are parallel and are not the edges of a single face. List pairs of opposite edges. How many are there?

   b) Two vertices of a cube are called opposite if they are not joined by an edge and are not the vertices of a single face. List pairs of opposite vertices. How many are there?

3. In the cube pictured in Exercise 2 vertices *B, D, E, G* are the vertices of a tetrahedron inscribed in the cube.

   a) Sketch a cube (in one color) and draw in the edges of an inscribed tetrahedron (in another color).

   b) Can you find additional inscribed tetrahedra? How many?

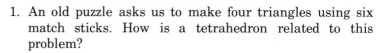

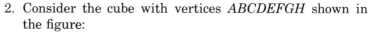

4. Use a model (if available) or a photograph to count the number of edges for each of the regular convex polyhedra. Count the number of vertices for each and the number of faces.

	Number of edges (E)	Number of vertices (V)	Number of faces (F)
Tetrahedron			
Cube			
Octahedron			
Dodecahedron			
Icosahedron			

5. After completing the table in Exercise 4, find the relationship between $V + F$ and E. This relationship is called **Euler's Formula.**

6. Since each face of the cube contains *four* edges, and *three* edges meet at each vertex, we use a symbolism suggested by Schläfli to describe the cube. Specifically, the cube is described by (4,3). Complete the table to describe the other polyhedra.

Regular polyhedron	Schläfli symbol
Cube	$(4, 3)$
Tetrahedron	
Octahedron	
Dodecahedron	
Icosahedron	

7. When two-dimensional pictures are made of three-dimensional solids (as in Exercise 2), dashed lines are used to show edges that are hidden from view. Practice making accurate drawings of a cube, a tetrahedron, and an octahedron. Be sure to include the dashed lines.

In one first-grade class the students counted the faces of a large cardboard model of a dodecahedron by having each of 12 pupils place one hand on each of the 12 faces (Note this concrete application of one-to-one correspondence).

8. A pattern that can be cut out and fastened together to form a polyhedron is called a net for the polyhedron. Figure 4.15 shows a net for each of the regular polyhedra.

   a) Make an appropriate-sized net and construct a model of one of the regular polyhedra.

   b) Draw three different nets of six squares that can be used to form a cube.

   c) Draw as many different nets of four equilateral triangles as you can which will form a tetrahedron.

9. If a cube is drawn and a point is marked at the center of each face of the cube, and these points are connected, an octahedron is formed. We say that the *dual* of a cube is an octahedron. More generally, the dual of any convex regular polyhedron is another convex regular polyhedron formed by connecting the centers of the faces of the first polyhedron.

   a) List each of the five regular polyhedra and find the dual of each.

   b) Compare your answers in part (a) with the table you made in Exercise 2.

10. Suppose a wire model of a polyhedron is made showing just the edges of the polyhedron. If the model is viewed in perspective from a position just outside the center of one face, this face appears as a frame and all the remaining

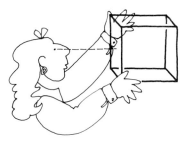
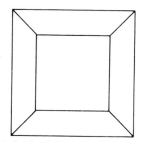

A Schlegel diagram for a cube

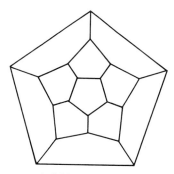

A Schlegel diagram
for a regular dodecahedron

edges are seen in its interior. If this view is drawn for a solid, a diagram called a Schlegel diagram is formed. Draw a Schlegel diagram for a tetrahedron and an octahedron.

11. A **deltahedron** is a solid with equilateral triangles as faces. The tetrahedron, with four equilateral triangular faces, is the smallest deltahedron. The icosahedron, with twenty equilateral triangular faces, is the largest deltahedron.

   a) Which other regular polyhedron is also a deltahedron? How many faces does it have?

   b) The deltahedra listed so far suggest that each deltahedron has an even number of faces. Use the method of Investigation 4.2 to construct deltahedra of twelve and fourteen faces.

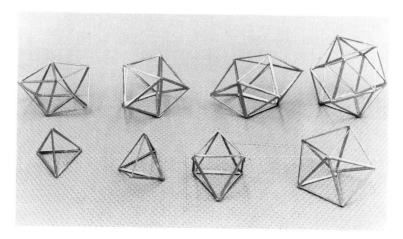

Models of the eight deltahedra

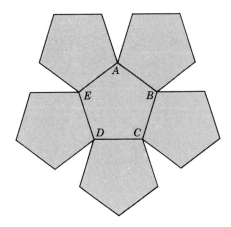

12. This exercise outlines a quick way of constructing a dodecahedron.

   a) Cut from poster board two nets like the one shown
at the left.

   b) Crease sharply along edges of *ABCDE*.

   c) Place one net upon the other, one rotated 36°,
and weave an elastic band alternately above and below
the corners.

   d) Raise your hand and watch the dodecahedron pop up.

Construct a dodecahedron using this method.

13. This exercise outlines a method of constructing a tetrahedron from a sealed envelope.

   a) Construct point C so that $\triangle ABC$ is an equilateral
triangle.

   b) Cut along $\overline{DE}$, through C, and parallel to AB.

   c) Fold along $\overline{AC}$ and $\overline{BC}$ back and forth in both directions.

   d) Let C' be the point on the reverse side corresponding
to point C.

   e) Pinch the envelope so that points D and E are joined
and C and C' are separated. Tape along segment CC'
and the tetrahedron is complete.

Construct a tetrahedron using this method.

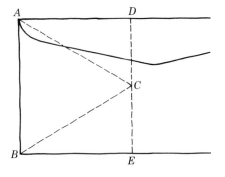

14. Puzzles

a) Can you make two of these solids and place them together to form a tetrahedron?

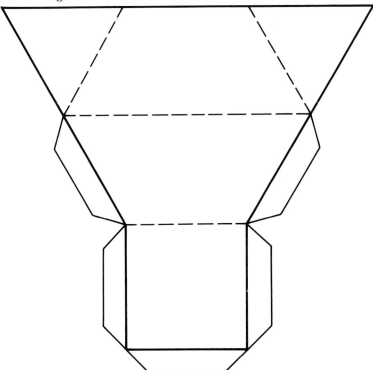

Many students at various ages need hands-on experiences in order to develop the concepts of solid geometry. Building and investigating polyhedra made from straws and pipe cleaners can provide this experience. Mary Laycock's book called *Straw Polyhedra* (Creative Publications, 1970) is helpful in this regard.

b) Can you make three of these solids and place them together to form a cube?

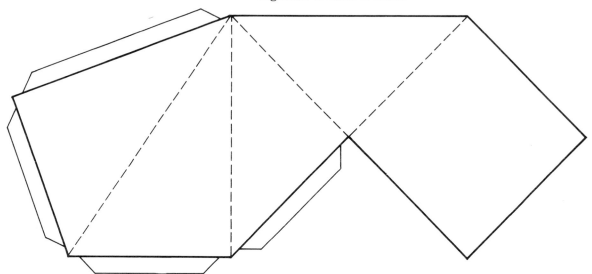

c) Can you make two of these solids and put them together to form a cube?

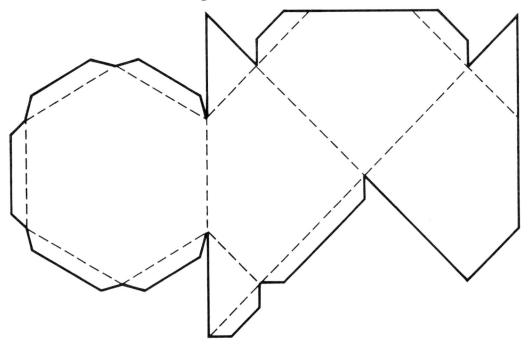

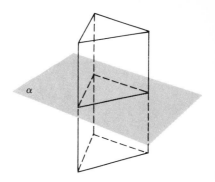

Figure 4.16

SYMMETRY IN SPACE

From pre-Egyptian times, through the age of Theaetetus and Euclid, and into the twentieth century, the desire to find order and regularity in nature and in one's creations has stimulated tremendous activity and accomplishment. Herman Weyl, in his beautiful book *Symmetry* asserts:

> Symmetry, as wide and narrow as you may define its meaning, is one idea by which man through the ages has tried to comprehend and create order, beauty, and perfection.

For example, symmetry is an important theme in the study of molecular structure. Diamonds and sodium chloride are but a few of the crystalline structures possessing symmetry.

Even in the simple task of getting better acquainted with polyhedra, it is instructive to use the idea of symmetry.

In Fig. 4.16, α is called a *plane of symmetry* of the equilateral triangular prism, while in Fig. 4.17 β is not a plane of symmetry for the triangular pyramid.

Until a careful definition is given in a later chapter, it will suffice to provide an intuitive test to determine whether or not a given plane is a plane of symmetry for a polyhedra.

One such test is the "mirror test." If the plane in question, say α in Fig 4.16, were a mirror and the part of the prism behind this mirror were removed, the observer could look in the mirror and see the original, complete prism. That is, one half of the prism is the mirror image of the other half. Hence the plane α would qualify as a plane of symmetry of the prism. On the other hand, if the plane in question, say β in Fig. 4.17, were a mirror and the part of the pyramid on the lower side of this mirror were removed, the observer could not see the original, complete pyramid when he looked in the mirror. Instead he would see a six-faced deltahedra as pictured in Fig. 4.3. Hence the plane β is not a plane of symmetry for the pyramid.

If a plane of symmetry exists for a polyhedron, we say the polyhedron has **reflectional symmetry**. The following investigation provides an opportunity to explore planes of symmetry for selected polyhedra.

One important goal in teaching geometry is to help students develop their space perception. If your school art department has a hot wire for cutting styrofoam, the styrofoam polyhedra available from Cuisenaire Co. can be sliced along a plane of symmetry and a mirror can be used to illustrate the idea of a plane of symmetry.

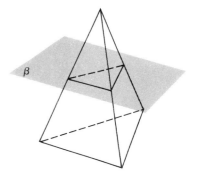

Figure 4.17

Investigation 4.3

A. Use models to count the total number of planes of symmetry for the tetrahedron, cube, and octahedron by using *one* of the following techniques.

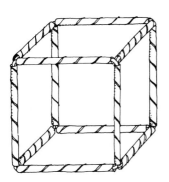

   1. Construct the models from drinking straws and pipe cleaners, and identify symmetry planes with 4 x 6 file cards glued into place.

   2. Construct the models from nets and visualize the symmetry planes.

   3. Use commercially produced models and attempt to visualize the symmetry planes. (Styrofoam models are helpful since they can be cut into pieces to illustrate planes of symmetry.)

B. Reflect on part A. Did you devise a systematic method of counting the symmetry planes? Compare your methods with others in your class.

C. Use the method your class has decided is most efficient and count the number of symmetry planes for the dodecahedron and the icosahedron.

In addition to reflectional symmetry, a consideration of cyclic, or rotational, symmetry often supplies one with a fresh new view of a solid.

For example, if this flower is rotated 72° about an axis perpendicular to its plane and through its center, its original appearance is reproduced. Since one can rotate the flower through 72° five times before it actually is returned to its original position, we say that the flower has **rotational symmetry of order 5** about the **axis of rotational symmetry** described above. We also say the flower has **72° rotational symmetry.**

Walter Dawn

Patty Benner

While flowers frequently have rotational symmetry of order 5, their winter season counterpart, the snowflake, invariably has rotational symmetry of order 6, since a rotation of a snowflake through an angle of 60° about an axis through its center and perpendicular to its plane reproduces its original appearance.

When we consider polyhedra, we note that in Fig. 4.18, line l is an axis of rotational symmetry of the square pyramid while line r is not an axis of rotational symmetry of the cuboid.

As with reflectional symmetry, we will defer a careful definition of rotational symmetry to a later chapter, and present at this stage an intuitive test for deciding whether or not a polyhedron has an axis of rotational symmetry.

If one thinks of a line in question, say line l in Fig. 4.18, as a rod or a straw and of the pyramid as a model, the person could rotate the model (through an angle other than 360°) so that the pyramid appears to be in the same position as before the rotation. Hence the line l would qualify as an axis of rotational symmetry of the pyramid. Since the pyramid can be rotated through an angle of 90° four times before it returns to its actual original position, the pyramid has rotational symmetry of order 4, or 90° rotational symmetry about this axis.

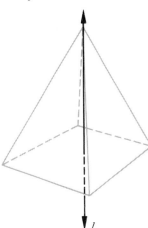

Figure 4.18

On the other hand, if line r is considered and one uses models, the cuboid (rectangular parallelepiped) model cannot be rotated about the rod r (through an angle other than 360°) so as to appear in the same position as before the rotation. Hence r is not an axis of rotational symmetry of the cuboid.

The following investigation will provide an opportunity for you to search for axes of rotational symmetry for selected polyhedra.

Investigation 4.4

A. Use a paper net or an envelope to make a model of a tetrahedron. Cut off a vertex, cut a small slot in the center of one face, and insert a drinking straw as shown in the figure.

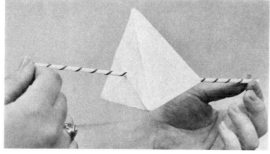

The straw represents an axis of rotational symmetry of the tetrahedron. The tetrahedron has rotational symmetry of what order about this axis?

Can you cut slots and use a straw to show each of the seven axes of rotational symmetry of the tetrahedron?

B. Use a paper net, make a model of a cube, cut slots, and use a straw to show each of the axes of rotational symmetry of the cube. How many axes of rotational symmetry does a cube have?

C. Use a paper model and drinking straws to determine the number of axes of rotational symmetry of an octahedron.

In Investigation 4.3, the analysis of the planes of symmetry of the tetrahedron should have established that such a plane contains an edge of a tetrahedron and the midpoint of an opposite edge. (See Fig. 4.19.) Since a tetrahedron has six edges, it has six different planes of symmetry.

In Investigation 4.4, the analysis of the axes of rotational symmetry of the tetrahedron should have revealed that there are several types of axes of symmetry.

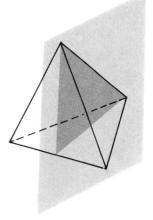

Figure 4.19

One type passes through one vertex and the center of a face opposite this vertex, as shown in Fig. 4.20(a). Since the regular tetrahedron has four vertices, it has four such axes. Since the face is an equilateral triangle, the order of rotational symmetry about each of these axes is three.

Another type of axis of symmetry for the tetrahedron passes through the midpoints of two opposite edges, as in Fig. 4.20(b). Since the tetrahedron has three different pairs of opposite edges, it has three such axes of rotational symmetry. Since a rotation of the tetrahedron through 180° about one of these axes returns it to a position that appears the same as before the rotation, the order of rotational symmetry about each of these axes is 2.

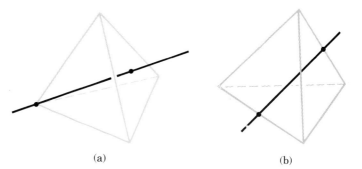

Figure 4.20   (a)   (b)

EXERCISE SET 4.2

1. Draw pictures of the following three-dimensional figures and show one plane of symmetry.

    a) cube   b) tetrahedron

    c) octahedron   d) square pyramid

    e) cuboid   f) pentagonal prism

    g) cylinder   h) hexagonal prism

2. a) Is the plane through △ABC a plane of symmetry of the cube?

    b) How is the △ABC related to Exercise 3 on page 120?

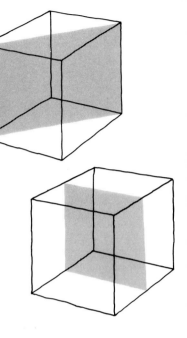

3. a) A plane passing through opposite edges of a cube is a plane of symmetry for the cube. How many planes of symmetry of this type are there for a cube?

   b) A plane midway between a pair of opposite faces of a cube is a plane of symmetry for the cube. How many planes of symmetry of this type are there for a cube?

4. There are three types of axes of symmetry for a cube.

   a) How many axes of symmetry of order 3 are there for a cube?

   b) How many axes of symmetry of order 2 are there for a cube?

   c) How many axes of symmetry of order 4 are there for a cube?

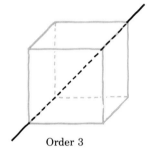
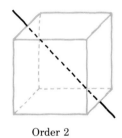
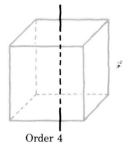

Order 3

A line through a
pair of opposite
vertices

Order 2

A line through
the midpoints of
opposite edges

Order 4

A line through
the centers of
opposite faces

5. How many planes of symmetry does each of these polyhedra have?

   a) a square pyramid

   b) a pentagonal prism

   c) a cuboid

   d) a cone

   e) an octagonal prism

   f) an isosceles triangular prism

6. Complete this table. The strategy outlined in Exercises 3 and 4 may help.

| Polyhedron | Number of axes of rotational symmetry | | | | | | Number of planes of symmetry |
	of order 2	of order 3	of order 4	of order 5	of order 6	Total	
Regular tetrahedron							
Equilateral triangular prism							
Square pyramid							
Cube							
Regular octahedron							

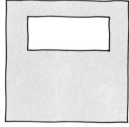

(a)

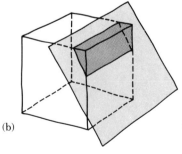

(b)

7. When an index card with a hole cut in it (a) is placed over a cube (b) so that it "just fits," a model of a plane intersecting a cube is produced. The intersection of a cube and a plane is called a *cross section* of a cube.

Use index cards and a model of a cube. Cut a hole in an index card that fits on your cube to show each of these cross sections of the cube.

a) square

b) isosceles triangle

c) equilateral triangle

d) trapezoid

e) parallelogram

f) pentagon

*g) regular hexagon

8. How many axes of rotational symmetry does a cuboid have?

*9. A cube has numbers associated with each corner as shown in the figure. In how many different ways may the cube be placed in the box?

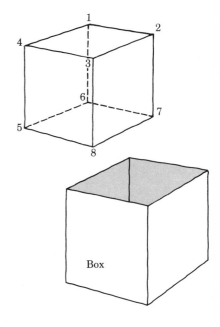

Box

*10. Which of the following objects possess neither reflectional nor rotational symmetry?

a) coffee cup

b) bowling ball

c) hand saw

d) French horn

e) scissors

f) table knife

*SEMIREGULAR POLYHEDRA

In Chapter 4 we studied both regular tessellations and semi-regular tessellations. Recall that regular tessellations were constructed by using regular polygons of one type, whereas semiregular tessellations were constructed by using regular polygons of more than one type. In addition, all vertices of semiregular tessellations are surrounded in the same way. More simply, we say that all vertex figures of a semiregular tessellation are alike.

There are similar differences between the regular and the semiregular polyhedra. A semiregular polyhedron is a polyhedron constructed from regular polygons of more than one type such that all vertex figures are alike. For example, in Fig. 4.21 we see a polyhedron with regular octagon and equi-lateral triangular faces with isosceles triangles for vertex figures. This polyhedron is one of the fourteen semiregular polyhedra.

Several of the semiregular polyhedra can be obtained from the regular polyhedra through a process known as trun-cating, or "slicing off the vertices." In the case of a cube, if the plane of this slice is perpendicular to an axis of symmetry through the vertex, the new face resulting from the trunca-

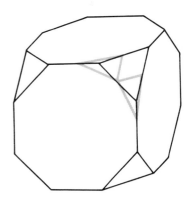

Figure 4.21

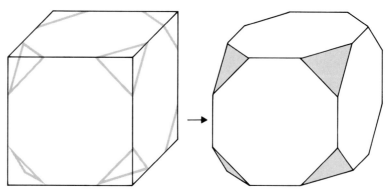

Figure 4.22

Truncated cube

tion is an equilateral triangle. If all eight vertices are truncated so that the original square faces of the cube become regular octagons, the new polyhedron is called a **truncated cube**. Figure 4.22 illustrates that the polyhedron of Fig. 4.21 is a truncated cube.

Similarly, as in Fig. 4.23, if we truncate each vertex of the octahedron (the dual of the cube) so that the original equilateral faces become regular hexagons after the truncation, we produce a **truncated octahedron**.

A third solid, which we might think of as lying between the truncated octahedron and the truncated cube, is obtained by dividing each edge of *either the cube or the octahedron* into halves, connecting the points as shown in Fig. 4.24, and removing the corners. It is called a **cube octahedron**. We can also view this as a truncation of the vertices by planes which pass through the midpoints of the edges.

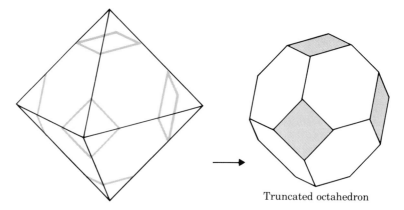

Figure 4.23

Truncated octahedron

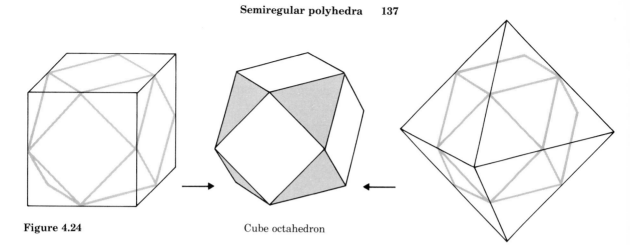

Figure 4.24

Cube octahedron

So from the cube and octahedron we obtain three semiregular polyhedra, pictured in Fig. 4.25.

Figure 4.25

Many opportunities exist for older students to become involved with building models of semiregular polyhedra. An analysis of the Euler characteristic and of the lines and planes of symmetry of these polyhedra can often be quite challenging. The book *Mathematical Models* (Oxford University Press, 1961) by Cundy and Rollett is an excellent aid. Also, *Shapes, Spaces, and Symmetry* (Columbia University Press, 1971) by Allan Holden has an excellent description of a method for making polyhedra models.

There are two semiregular polyhedra which we can imagine as being obtained from the cube octahedron through a truncation followed by a slight distortion. We see in Fig. 4.26 that each vertex of a cube octahedron can be truncated so that the equilateral triangular faces become regular hexagons and the square faces become regular octagons. This truncation transforms the vertices into rectangles which we imagine are distorted into squares. This new polyhedron carries two names—the **truncated cube octahedron** or **great rhombicuboctahedron**. The second polyhedron obtained from the cube octahedron is called the **rhombicuboctahedron**. We find the midpoints of the edges of the cube octahedron and connect them as shown in Fig 4.27. Cutting off the vertices along the shaded lines yields a polyhedra with square faces, equilateral triangular faces, and rectangular faces. Again, imagine these rectangular faces are distorted into squares. The resulting polyhedron is the rhombicuboctahedron.

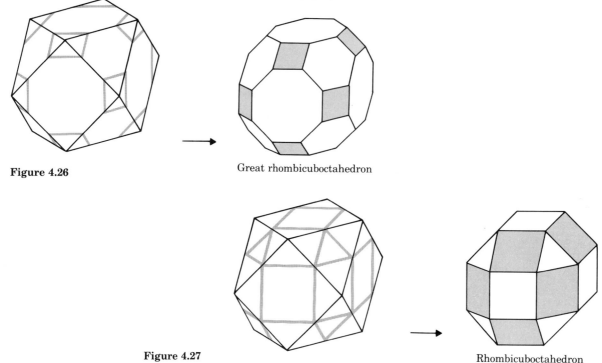

Figure 4.26

Great rhombicuboctahedron

Figure 4.27

Rhombicuboctahedron

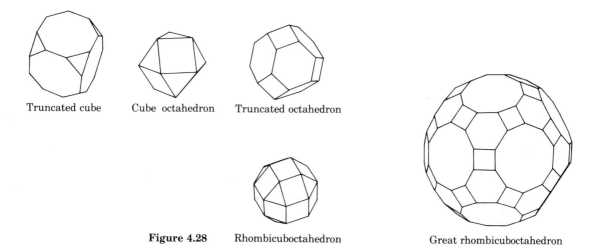

Truncated cube   Cube octahedron   Truncated octahedron

Figure 4.28   Rhombicuboctahedron   Great rhombicuboctahedron

So we see that through the truncation process we obtain, beginning with the cube and the octahedron, five semiregular polyhedra—as pictured in Fig. 4.28 and listed in Table 4.1.

Table 4.1

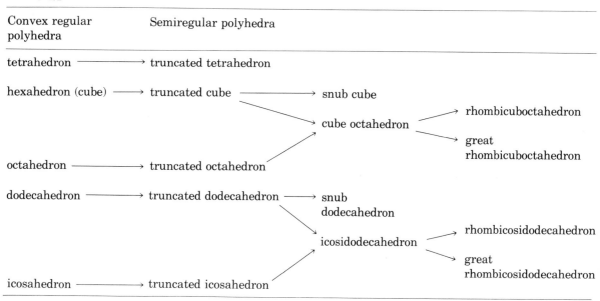

Figure 4.29

Truncating the icosahedron and the dodecahedron we obtain the truncated icosahedron and truncated dodecahedron with the icosadodecahedron midway between these two playing a role analogous to the cube octahedron. Truncation followed by distortion applied to the icosadodecahedron yields the **great rhombicosidodecahedron** and the **rhombicosidodecahedron.** So the dodecahedron and the icosahedron yield five more semiregular polyhedra as pictured in Fig. 4.30.

The three remaining polyhedra listed in Table 4.1 are the **truncated tetrahedron,**[*] the **snub cube,**[†] and the **snub dodecahedron**[‡] which are pictured in Fig. 4.31. These thirteen semiregular polyhedron were known to Archimedes and are hence often referred to as the Archimedean polyhedra.

---

[*] The truncated tetrahedron is formed by dividing edges of the tetrahedron into three parts, connecting the points, and removing the corners.

[†] The snub cube can be produced by drawing one diagonal in each of the twelve squares of the rhombicuboctahedron that correspond to the edges of the cube, and then distorting each of the right triangles thus produced into an equilateral triangle. The diagonals are chosen so that just one diagonal passes through each of the 24 vertices.

[‡] The snub dodecahedron can be reproduced in a manner similar to that of the snub cube by drawing appropriate diagonals on each of the 30 squares of the rhombicosidodecahedron.

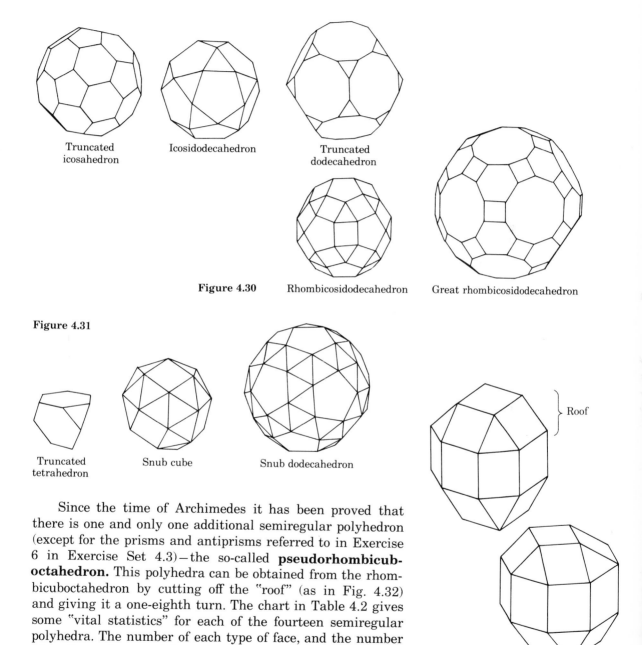

Truncated
icosahedron

Icosidodecahedron

Truncated
dodecahedron

Figure 4.30   Rhombicosidodecahedron

Great rhombicosidodecahedron

Figure 4.31

Truncated
tetrahedron

Snub cube

Snub dodecahedron

} Roof

Since the time of Archimedes it has been proved that there is one and only one additional semiregular polyhedron (except for the prisms and antiprisms referred to in Exercise 6 in Exercise Set 4.3)—the so-called **pseudorhombicuboctahedron.** This polyhedra can be obtained from the rhombicuboctahedron by cutting off the "roof" (as in Fig. 4.32) and giving it a one-eighth turn. The chart in Table 4.2 gives some "vital statistics" for each of the fourteen semiregular polyhedra. The number of each type of face, and the number of vertices, edges, and faces are listed.

Figure 4.32

Table 4.2

Polyhedron	Triangles	Squares	Pentagons	Hexagons	Octagons	Decagons	V	E	F
Tetrahedron	4	—	—	—	—	—	4	6	4
Hexahedron	—	6	—	—	—	—	8	12	6
Octahedron	8	—	—	—	—	—	6	12	8
Dodecahedron	—	—	12	—	—	—	20	30	12
Icosahedron	20	—	—	—	—	—	12	30	20
Tr. tetrahedron	4	—	—	4	—	—	12	18	8
Tr. cube	8	—	—	—	6	—	24	36	14
Tr. octahedron	—	6	—	8	—	—			
Tr. dodecahedron	20	—	—	—	—	12			
Tr. icosahedron	—	—	12	20	—	—	60	90	32
Cube octahedron	8	6	—	—	—	—	12	24	14
Rhombicuboctahedron	8	18	—	—	—	—	24	48	26
Pseudorhombicubocta.	8	18	—	—	—	—	24	48	26
Tr. cuboctahedron	—	12	—	8	6	—	48	72	26
Snub cube	32	6	—	—	—	—	24	60	38
Icosidodecahedron	20	—	12	—	—	—	30	60	32
Rhombicosidodecahedron	20	30	12	—	—	—	60	120	62
Tr. icosidodecahedron	—	30	—	20	—	12	120	180	62
Snub dodecahedron	80	—	12	—	—	—	60	150	92

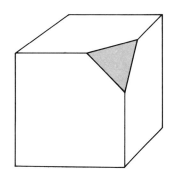

EXERCISE SET 4.3

1. a) Suppose one corner of a cube is truncated. Describe the axes of rotational symmetry of this figure. Describe the planes of reflectional symmetry.

   b) Suppose two corners of the cube are truncated. Answer the questions in part (a) for this figure.

2. Does Euler's Formula (Exercise 5, Exercise Set 4.1) hold for semiregular polyhedra?

3. In a cube octahedron each vertex is surrounded by a triangle, square, triangle, square in that order. Consequently, this polyhedron is often represented with the notation (3,4,3,4). Write the parentheses notation for each of the semiregular polyhedra.

4. When a cube is truncated, each of the vertices of the cube becomes 3 vertices of the truncated cube. Consequently, there are 8·3 = 24 vertices of a truncated cube. Likewise,

three new edges are added to the 12 edges of the cube when a truncated cube is formed.

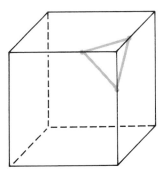

Use this counting strategy to find the number of vertices, edges, and faces for the truncated octahedron and the truncated dodecahedron.

5. a) Sketch a prism with regular hexagon bases and square sides. Do the same for a prism with regular octagons. Are these polyhedra semiregular?

    b) How many prisms are there which are semiregular polyhedra?

6. An **antiprism** is a prism-type solid with triangular rather than square sides.

    a) Construct with rubber bands and poster board an antiprism that is a semiregular polyhedron.

    *b) How many different such antiprisms are there?

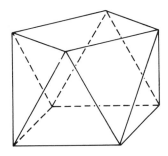

7. Put four of the solids from the puzzle in Exercise 14(c), Exercise Set 4.1 together to form a truncated octahedron.

*TESSELLATIONS OF SPACE

In Investigation 4.1 the experience with wall building suggested that the cuboid could be used to "fill" space. That is, in much the same way as we considered polygonal tessellations of the plane in Chapter 3, we can consider polyhedral tessellations of space. Obviously, the cube can also be used to tessellate space. The simple packing of boxes into larger cartons, which occurs every day, illustrates this packing for these polyhedra. Should we desire a more artistic packing or should our curiosity be aroused, a resulting natural question is, What other polyhedra can be used to tessellate space? If there are models available to you, completing the following investigation may help answer this question.

Investigation 4.5

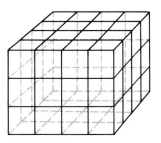

A. Since a square will tessellate the plane, a right square prism (cuboid) will tessellate space. Can you use your knowledge of plane tessellations and experiment with models to decide which other prisms will tessellate space?

B. The cube is a regular polyhedron which tessellates space. Use models of the other regular polyhedra and decide which ones will fill space. Are there any combinations of regular polyhedra which will fill space? Make careful observations and be ready to support your conclusions.

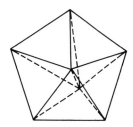

Figure 4.33

In deciding whether or not regular tetrahedra can be used to fill space, the experimenter often tries to fit five tetrahedra about a common edge, as shown in Fig. 4.33. Upon seeing that these five tetrahedra seem to fit, the observer assumes that clusters of five polyhedra like this can be fitted together to tessellate space.

Clearly the five tetrahedra will fit around an edge only if the dihedral angle of a tetrahedron is 1/5 of 360°.

To find the degree measure of this angle, we observe that six diagonals of the faces of a cube form the edges of a tetrahedron, as shown in Fig. 4.34. If we choose a cube in which the length of each side is 1, we notice that the length of BC is 1 and the length of AC is $\sqrt{2}/2$. It follows that:

$$\tan d = \sqrt{2}/2$$
$$\tan d = .707$$
$$d = 35°16'$$

Regular polyhedron	Degree measure of dihedral angle
Cube	90°
Tetrahedron	70°32′
Octahedron	109°28′
Dodecahedron	116°34′
Icosahedron	138°11′

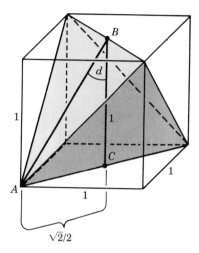

Figure 4.34

Thus the dihedral angle of a tetrahedron is $2 \times (35°16')$ or 70°32′. By using methods similar to the one illustrated in Fig. 4.34, the measures of the dihedral angles of the remaining regular convex polyhedra can be found. By checking multiples of the dihedral angles recorded in the table above, we see that the cube is the only one with a dihedral angle possessing a multiple equal to 360°. This observation suggests that the cube is the only regular polyhedron which tessellates space.

Are there combinations of regular polyhedra which tessellate space? Since $70°32' + 109°28' = 180°$, we see that two tetrahedra and two octahedra completely surround an edge. We suspect that a combination of tetrahedra and octahedra tessellate space.

To illustrate that this is true we consider space from two points of view. Figure 4.35 shows a way to place two tetrahedra and an octahedron together to make a **rhombohedron.** A rhombohedron is a cube which has been distorted by changing the faces from squares to rhombi, and like the cube will fill space. Hence combinations of octahedra and tetrahedra will tessellate space. (Note that $ABCD$ is one tetrahedron in Fig. 4.35, $PQRS$ is the other, and $QBRDCS$ is the octahedron.)

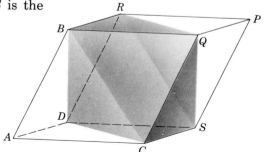

Figure 4.35

The second point of view comes by observing the tessellation of cubes in a different way. We have noted (Exercise 3, Exercise Set 4.1) that a tetrahedron may be inscribed in a cube as shown in Fig. 4.36.

If we observe a filling of space with cubes and focus on the inscribed tetrahedra, we observe in Fig. 4.37 that adjoining cubes contribute eight right pyramids like *ABCD* in Fig. 4.36 to form one octahedron surrounding each vertex of the cube tessellation not used by the tetrahedra. Thus the cube tessellation in proper perspective yields the tessellation of *octrahedra* and *tetrahedra*. Each edge in this tessellation is surrounded alike.

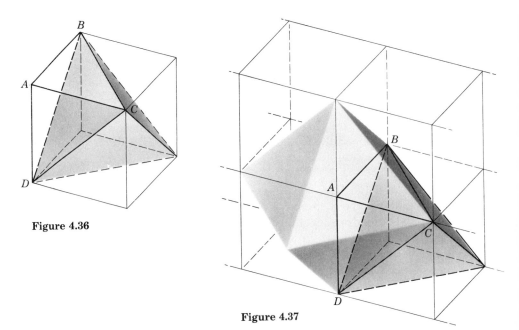

Figure 4.36

Figure 4.37

Still other tessellations are revealed by viewing the tessellation of cubes from the proper perspective. For example, a cube can be cut into a cube octahedron and eight right pyramids (like *ABCD* in Fig. 4.38). If we consider a filling of space with cubes and focus on the inscribed cube octahedra, we observe in Fig. 4.39 that adjoining cubes contribute eight right

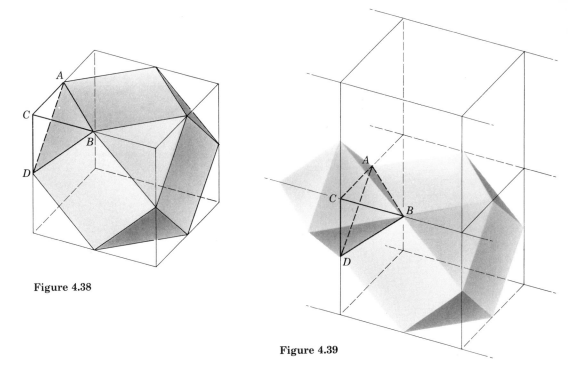

Figure 4.38

Figure 4.39

pyramids like *ABCD* in Fig. 4.38 to form an octahedron surrounding each vertex of the cube tessellation. Thus the cube
tessellation in the proper perspective (Fig. 4.39) yields the
tessellation of space by cube octahedra and octahedra.

   In working with the puzzle in Exercise 14, Exercise Set
4.1, you may have observed that a cube can be "sliced" by a
plane that is the perpendicular bisector of the axis joining opposite vertices to form two parts with regular hexagonal faces
(Fig. 4.40).

   If we view a filling of space with cubes and focus on these
halves, we observe that eight adjacent cubes contribute sufficient halves to form truncated octahedron (Fig. 4.41). The
remaining halves can be arranged to contribute to other adjacent truncated octahedra which together with the original
completely fill space. Thus the cube tessellation in the proper
perspective yields the tessellation of space by the *truncated
octahedron*. If this tessellation is transformed so that the 14-
faced truncated octahedra become bricks, we arrive at the
arrangement of Fig. 4.1 shown earlier in this chapter.

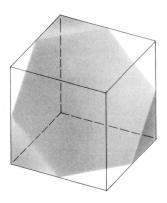

Figure 4.40

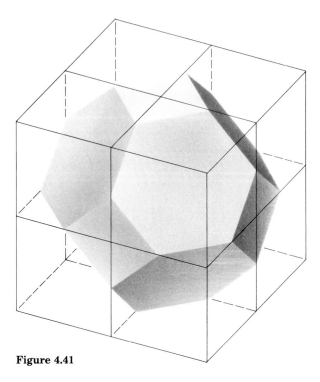

Figure 4.41

The four tessellations described above are all characterized by the property that in a given tessellation, all edges (and vertices) are surrounded alike.

There is only one more tessellation of space that has this property, namely the tessellation with a *tetrahedron* and three *truncated tetrahedra* around each edge. This tessellation may also be viewed by considering the tessellation of cubes in a certain way, but is more difficult to picture than the other four.

Thus we have the five possible tessellations of space with regular or Archimedean polyhedra in which each edge (and vertex) is surrounded alike.

1. Cube (four cubes around each edge)

2. Truncated octahedra (three around each edge)

3. Tetrahedra and octahedra (two of each around each edge arranged alternately)

4. Cube octahedra and octahedra (one octahedron and two cube octahedra around each edge)

5. Tetrahedron and truncated tetrahedron (one tetrahedron and three truncated tetrahedra around each edge)

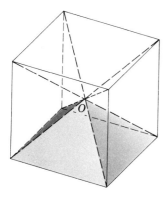

Figure 4.42

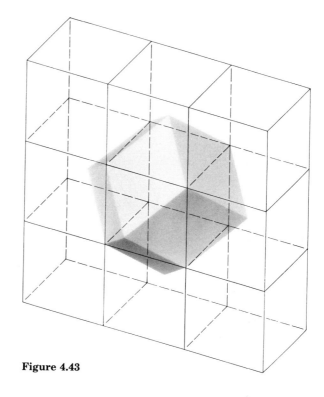

Figure 4.43

Still another interesting tessellation of space can be described by again considering the basic tessellation of cubes. Note that each cube can be viewed as six square-based pyramids with bases that are the faces of the cube and with vertices meeting at the center O of the cube as in Fig. 4.42.

If we consider the filling of space with cubes and view every other cube as being formed by six of the pyramids described in Fig. 4.42, we observe that each of the other cubes can "borrow" a pyramid for each of its faces (Fig. 4.43).

The polyhedron resulting from affixing a square pyramid to each face of a cube is called a **rhombic dodecahedron**. This solid has twelve rhombi as faces (see Fig. 4.44). It is not, however, an Archimedean polyhedron. Why?

Thus the cube tessellation in this particular perspective (Fig. 4.43), yields a tessellation of space by rhombic dodecahedra. Note that each vertex of this tessellation is not surrounded alike.

Figure 4.44

Cube with square-based pyramids affixed to each face.

Tessellation of rhombic dodecahedron described in Fig. 4.43.

From *Mathematical Snapshots* by H. Steinhause, 3rd edition. Copyright © 1950, 1960, 1968, 1969 by Oxford University Press, Inc. Reprinted by permission.

This last tessellation suggests that we can find other solids that fill space if we but allow ourselves to search beyond the regular and Archimedean polyhedra. If you continue this search you may find that space can be filled by such interesting figures as a *stellated rhombic dodecahedron,* a *rhombic tricontahedron,* a *trapezo-rhombic dodecahedron,* or even an *elongated rhombic dodecahedron.**

Clearly, as in the plane, the universe of space tessellation is rich and varied.

---

* Additional information about space tessellations can be found in W. W. Rouse Ball, *Mathematical Recreations and Essays,* New York: MacMillan, 1960 (2nd Ed.) Chapter 5 and Alan Holden, *Shapes, Space, and Symmetry,* New York: Columbia University Press, 1971.

EXERCISE SET 4.4

1. Since four four-sided polygons meet at each edge, the cube tessellation is denoted by 4^4. Since four equilateral triangles meet at each edge of the tessellation of octahedra and tetrahedra, it is denoted by 3^4. Write the symbol for the tessellations described in Figs. 4.37 and 4.39.

2. Use the information in the table below to verify that the sum of the dihedral angles around each edge in the five tessellations described on page 148 is 360°.

Polyhedron	Dihedral angles
Tetrahedron	70°32°
Octahedron	109°28′
Cube	90°
Cube octahedron	125°16′
Truncated octahedron	109°28′, 125°16′
Truncated tetrahedron	70°32′, 109°28′

3. Use any standard reference on polyhedra to find the dual of the rhombic dodecahedron.

BIBLIOGRAPHIC REFERENCES

[4, 22–29], [8, 129–139, 146], [11], [12], [13], [14, 6–40], [17, 5–13, 48–50], [20, 1–6, 92–96], [25, 148–157], [26], [28], [33], [34, 15–34, 65–74, 76–80, 161–180], [37], [39, 15–24, 91–118, 161–177], [43, 13–23], [51], [56], [64], [69], [81], [83, 82–88], [84], [91], [95], [98, 108–124, 154–169], [100, 102–123], [101], [103], [104], [105], [106], [107], [108]

PEDAGOGICAL ACTIVITIES FOR THE TEACHER

1. Analyze a basic textbook series for the elementary school to find the extent to which the program suggests experiences for the students with space figures. Make a list of the grade level and the page on which this introduction occurs. Suggest specific places where additional work with space figures might be included at a given level of your choice.

2. Suppose you were making out a requisition requesting the purchase of solid models for your classroom. Refer to current catalogs from mathematics laboratory materials suppliers and list the items you would purchase and what they would cost. Then suppose you have less than $50 dollars to spend. Which items would you buy?

3. Procure models of the five regular polyhedra and talk with primary school children about these solids. Are they interested? Can they count the number of faces? the number of edges? and the number of vertices? Can they think of objects in their world which remind them of the space figures? Write a summary of the children's reaction to the polyhedra.

4. Consult articles from the *Scientific American* (see bibliography at the end of this text), or a book such as *Mathematical Diversions* by Martin Gardner, and find at least two more puzzles which involve solid figures and are different from those in this chapter. Include these puzzles in your card file of geometric recreations.

5. A durable, attractively colored set of regular and Archimedean polyhedra are valuable materials to have in a classroom at any level. Refer to pages 188–192 in Alan Holden's *Shapes, Spaces, and Symmetry* for a method of making such models and make a set for your classroom. A carefully made set of models, even with extensive student handling, will last several years.

6. In order to encourage children to become actively involved in the exploration of a geometrical idea, an investigation card is often used. An example of such a card is shown here.

On a 5 by 5 geoboard, how many triangles with different shapes can you make that have no nails inside?

a) Use the following criteria to evaluate the above investigation card.

    i)   The objective(s) for the investigation card are clearly specified and appropriate.

    ii) The card has the potential to involve the student in an investigation of the ideas specified in the objectives.

    iii) The questions on the card are simple and clear, and motivate the student to become involved.

    iv) The card utilizes physical materials when appropriate and other materials that effectively aid in giving meaning to the ideas.

    v)   At least some part of the card has potential to create a situation in which a student is involved in sustained activity centered about the important idea.

    vi) The card allows possibilities for student choices both in investigation procedures and in modes of recording findings.

b) Create an investigation card of your own which would involve students at a given grade level in the exploration of a basic geometric idea. Try your card with students and try to improve its effectiveness.

5

Motions in the Physical World

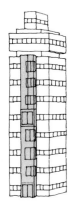

Small children have a primitive concept of motion and may perceive their mother moved to a different position as a different mother (*Scientific American*, 1971). Later they describe change of position of objects in terms of the end position only, disregarding the beginning position. Piaget considers the "change of position" developments in children in a chapter of *The Childs Conception of Geometry* (Harper and Row, 1964).

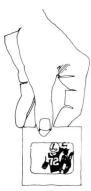

A glass elevator ride would ordinarily not be thought of as a "slide," but it is an example of this type of motion.

INTRODUCTION

Motion plays a significant role in the experience of persons of all ages. We live in a world of all kinds of motion and our perceptions of the properties of these motions are continually being refined and broadened. Because mathematics is useful in describing and analyzing motion, the mathematics of motion is an appropriate topic of study for students at various levels.

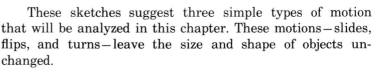

People have sought to develop perpetual motion machines for centuries. The picture at the right is a variation of Leonardo DaVinci's machine based on freely moving spheres.

These sketches suggest three simple types of motion that will be analyzed in this chapter. These motions—slides, flips, and turns—leave the size and shape of objects unchanged.

In Chapter 6 we shall study the geometrical relation of "congruence" by considering the mathematical abstractions of these motions, which we will call translations, rotations, and reflections.

The slide projector provides a useful example of the flip motion. When a slide-projected picture appears reversed on the screen, we remove the slide, flip it over, and place it back in the machine.

A ride on a ferris wheel is a vivid example of the turn motion.

SLIDES

A sliding motion is familiar to all of us. We experience it in our homes with sliding doors and sliding drawers; we have experienced it at play with the sled and the sliding board. We may have seen instances of how a job can be made a bit easier by sliding objects on an inclined plane.

In order to study the slide motion and its relationship to geometry more thoroughly we shall consider slides in the plane. Our geometric ideas will be suggested and reinforced by our experiences with physical objects. Consider, for example, placing a rectangular cutout exactly on the rectangle A.

✳ In measuring the length of a desk top with a foot ruler, the ruler must be slid from one position to another. Thus a primary school teacher might well introduce the ideas of slides (with focus on both beginning and ending points) before introducing the idea of measuring with a ruler.

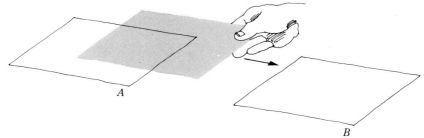

This cutout can be made to coincide with the rectangle *B* by simply *sliding* it along the paper without any turn or twist. We call rectangle *B* the **slide image** of rectangle *A*. The cutout was slid in a particular *direction* through a particular *distance*. Note that the before and after positions of the cutout determine this direction and distance.

The following investigation utilizes tessellations of the plane to provide a situation to help you develop your intuition about slides.

Investigation 5.1

Place tracing paper on this "tessellation" and trace it. If the tracing is slid (no turns or flips) from A to B, the tracing will "fit back on" the tessellation (assuming the original tessellation covers the complete plane, rather than just the portion shown).

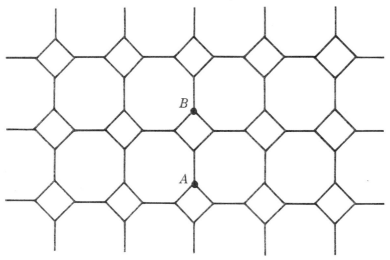

A. Starting with point A, find at least eight different *directions* along which the tracing can be slid to fit back on the tessellation. Describe a slide in each of these directions by marking a starting point and an ending point for the slide on the tracing paper.

Describe at least one more slide in a direction different from those already described which will enable the tracing to fit back on the tessellation.

B. For each tessellation (a,b,c), find at least six different directions along which the tracing can be slid to fit back on the tessellation. Give the starting and ending points to show one slide in each of these directions.

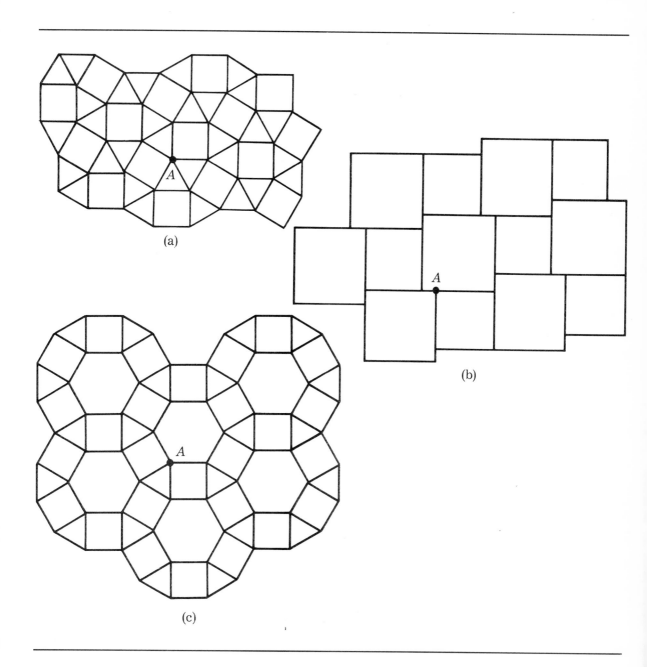

(a)

(b)

(c)

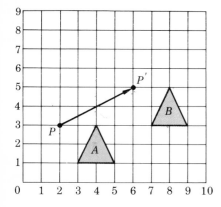

Figure 5.1

Since each slide has a certain direction and a certain distance, an arrow as PP' in Fig. 5.1 is often drawn to record or describe a certain slide. Since an arrow indicates both a specific *direction* and a specific *distance*, it completely determines a particular slide. The point P' (with coordinates $(6, 5)$) is the *slide image* of point P (with coordinates $(2, 3)$) after the slide PP'. We also say that triangle B is the slide image of triangle A after the slide PP'.

In Fig. 5.2 the slide images of the figures in black after the slide AA' are shown in gray. The reader should use tracing paper to show that this is true. Note that if A is marked on the tracing paper and the paper is slid so that A moves to A', then every other figure that has been traced on the tracing paper also moves through the distance and direction AA'.

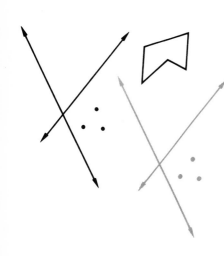

Figure 5.2

EXERCISE SET 5.1

1. In the sketches below, which of the figures on the right is a slide image of the figure on the left?

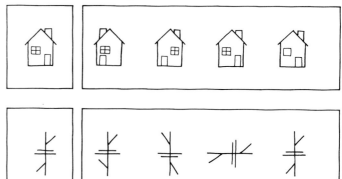

a)

b)

2. Draw triangle *ABC* and arrow *RS* on your paper. Lay a sheet of tracing paper over this paper and trace triangle *ABC* and point *R*. Slide the tracing paper so that the traced point *R* moves to point *S* on the original arrow. Push your pencil tip down on points *A*, *B*, and *C* so that impressions of these points show up on the underneath paper. Mark these impression points *A'*, *B'*, and *C'* and connect them to form the slide image of triangle *ABC* after the slide *RS*.

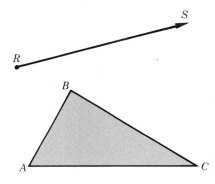

3. Draw quadrilateral *ABCD* and arrow *PQ* on your paper. Find the slide image of the quadrilateral after the slide *PQ*.

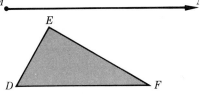

4. Line *l* is parallel to arrow *AA'*. Use line *l* and your compass to find the slide image of point *B* after the slide *AA'*.

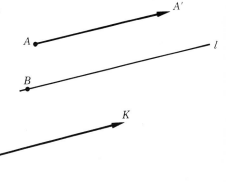

5. Draw arrow *JK* and point *X* on your paper. Construct a line through *X* which is parallel to a line through arrow *JK*. Use this line and your compass to find the slide image of point *X* after slide *JK*.

6. Use the idea in Exercise 4 three times to construct with ruler and compass the slide image of triangle *DEF* after slide *MN*.

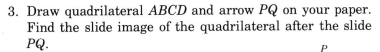

7. a) On graph paper draw five more arrows that describe the slide QQ'. (The arrows must have the same length and direction.)

   b) Give the coordinates of the slide image of points $A, B,$ and C after the slide QQ'.

   c) $\triangle J'K'L'$ is the slide image of $\triangle JKL$ after the slide EE'. Describe this slide by drawing an arrow for the slide on graph paper.

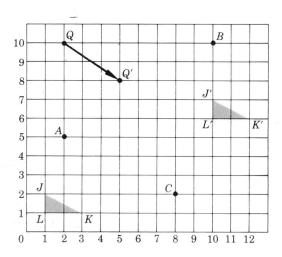

8. A student cut these "point sliders" from graph paper to describe three slides. The student actually used the cut-outs to find the slide images of certain figures.

   Use graph paper rather than tracing paper for the following exercises.

   a) The ends of a line segment AB have these coordinates: $A(3, 7)$ and $B(6, 4)$. Find the coordinates of the endpoints of the slide image of this segment after the slide pictured in (a).

   b) The vertices of triangle PQR have these coordinates: $P(2, 3)$, $Q(4, -2)$, and $R(-1, -3)$.

   Find the coordinates of the vertices of the slide image of this triangle after the slide pictured in (b).

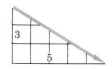

Down 3, right 5
"point slider"

(a)

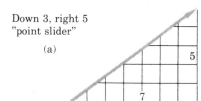

Right 7, up 5
"point slider"

(b)

c) Two points on line l have coordinates as follows: $R(3, 4)$ and $S(2, 6)$.

Find the coordinates of two points on the slide image of line l after the slide pictured in (c).

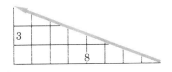

Left 8, up 3
"point slider"

(c)

9. Use geopaper (or something equivalent) for each of the following exercises.

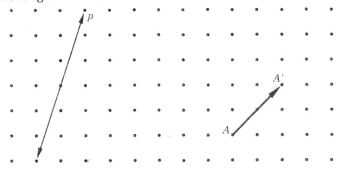

a) Draw the image of line p after slide AA'.

b) Draw a line q which slides onto itself after slide AA' (i.e., the slide image of q is q).

10. a) Does the image of line p under slide AA' pass through the point P?

b) Draw a line q whose image under slide AA' passes through point P.

c) Draw a second line q' whose image after the slide AA' passes through point P.

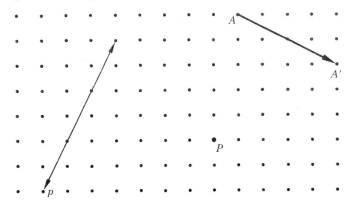

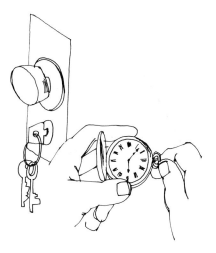

TURNS

A turning motion, just as a slide motion, is one which we have experienced since childhood and which we experience daily. We experience it in the door knob and the door key; we experience it when we roll up an automobile window, wind a watch, or wind a toy truck. We may have seen instances of how a job can be made easier by using a winding reel or a turning gear.

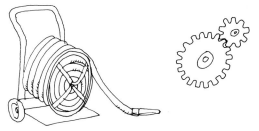

Our concept of the turning motion is refined and extended as we continue to experience turns not only in everyday life, but in the worlds of science and technology as well.

This Shinto symbol is symbolic of a turning motion and is said to represent the revolving universe.

In order to illustrate the turning motion using physical materials, consider two sheets of acetate fastened with a paper fastener. The $\triangle ABC$ is drawn on the underneath sheet and traced on the top sheet as shown in Fig. 5.3(a). If the underneath sheet is turned as shown in Fig. 5.3(b), triangle $A'B'C'$ can be traced on the top sheet.

We call $\triangle A'B'C'$ the **turn image** of $\triangle ABC$. The triangle was turned around point O (called the center of the turn) through a given angle (90°) which is called the **angle of the turn**.

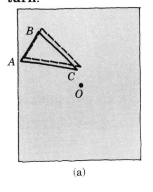

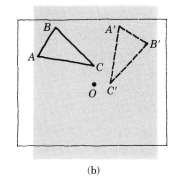

(a)

(b)

Figure 5.3

Our experience indicates that, perhaps because of the potential for hands-on activity it provides, the topic of "motions" is exciting for students of all ages. The booklets on *Motion Geometry* (Harper and Row, 1969) by Phillips and Zwoyer should be considered for elementary and junior high school children.

The following investigation utilizes tessellations of the plane to help you develop your intuition about turns.

"*You're fired!*"

Investigation 5.2

There are many interesting tessellation patterns in [63], like the tessellation shown in part b, which could be used in discussions of symmetry with older students.

Place tracing paper on this "tessellation" and trace it. If the tracing is held firmly at point F by the tip of a pencil and is rotated 180°, the tracing will "fit back on" the tessellation (assuming the original tessellation covers the complete plane, rather than just the portion shown).

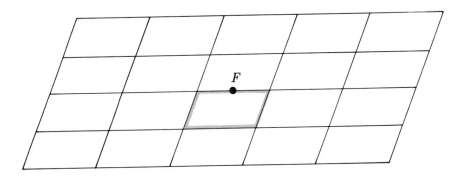

A. How may other points can you find on or inside the gray parallelogram for which this is true? Draw a portion of the tessellation and label these points.

B. If rotations other than 180° are allowed, how many points can you find?

C. Describe all possible centers for turns (such as point F) and the angles of the turns (such as 180°) that will make tracings of the following tessellations (a, b, c) fit back on themselves.

a)

b)

c)

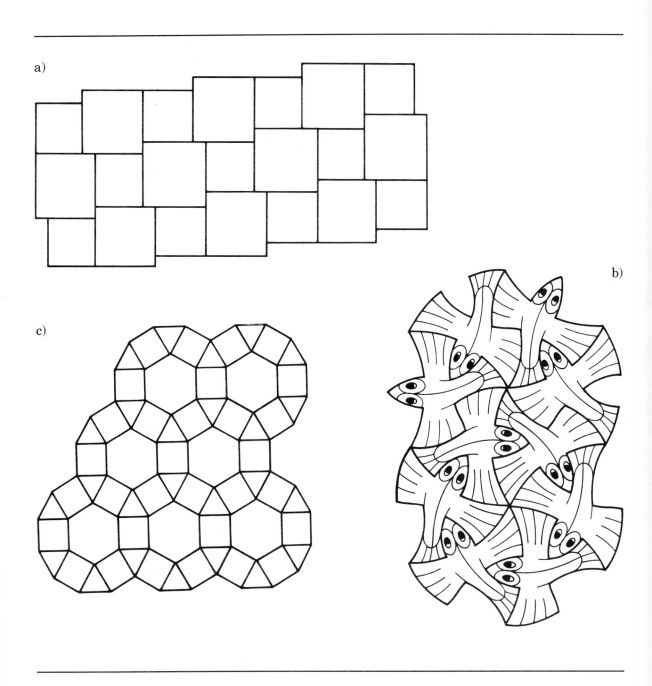

Since a turn is determined by specifying a center and an angle, the double arrow notation in Figs. 5.4 and 5.5 is used to describe turns. The notation in Fig. 5.4 indicates a 90° counterclockwise turn around point A, while a 180° clockwise turn about point B is indicated in Fig. 5.5.

Figure 5.4

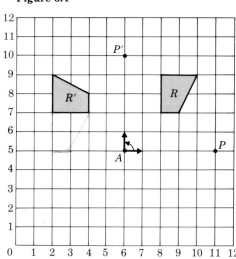

Figure 5.5

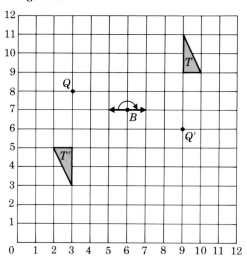

In Fig. 5.4 point P' is the 90° counterclockwise *turn image* of point P around point A. Note that distance AP is equal to distance AP'. We also say that quadrilateral R' is the 90° counterclockwise turn image of quadrilateral R around point A. In Fig. 5.5 point Q' is the 180° clockwise turn image of point Q around point B. We also say that T' is the 180° clockwise turn image of T around point B.

The figures in gray in Fig. 5.6 are the 60° clockwise turn images around the point O of the figures in black. The reader should use tracing paper to show that this is true. Note that if tracing paper is held stationary at point O with the tip of a

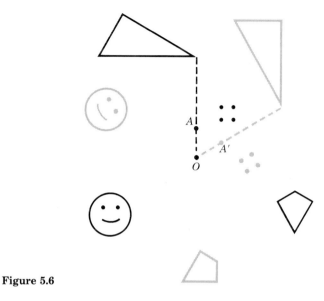

Figure 5.6

pencil and the tracing of point *A* is turned to fit upon point *A'*, then every other figure on the tracing paper is turned through an angle of 60° around point *O*.

EXERCISE SET 5.2

1. The situations for this exercise are taken from the book *The Unexpected Hanging* by Martin Gardner (New York: Simon & Schuster, 1969).

Turn this card 90° counterclockwise. What do you see?

NOW NO
SWIMS
ON MON

A sign like this appeared beside a public swimming pool. Turn this card 180°. What do you notice?

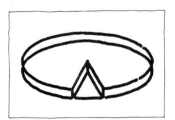

A slice is missing from this cake. Can you turn the card in such a way as to find the slice?

337-31770

Oliver Lee asked that the above number appear on his license plate. Turn the card 180°. Can you explain why he made this request?

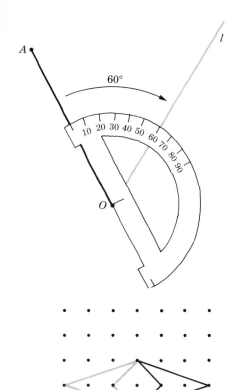

2. Mark a point O (the center of a turn) and a point A as shown in this picture on your paper. Draw OA and use your protractor as shown to draw a line (l) forming an angle of 60°.

   a) Use tracing paper to find the 60° clockwise turn image of point A around point O. Label this point A'.

   b) Use your compass to find the image point described in part (a).

3. Draw a triangle ABC and a point O on your paper.

   Use a protractor and tracing paper to find the 90° counterclockwise turn image of triangle ABC about point O.

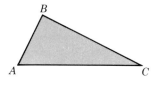

4. In each of the figures shown, the gray figure is the image of the black figure under a turn. Label the center of the turn and indicate the angle measure for each turn. (Tracing paper may be helpful.)

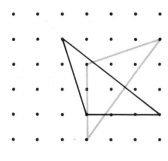

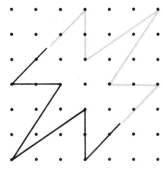

5. Give the coordinates of the point that is the 90° counterclockwise turn image of point A around point O.

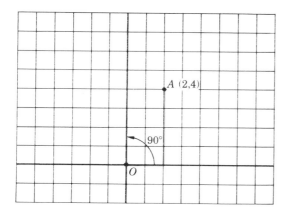

6. Draw this quadrilateral on graph paper. Draw in red the
   90° counterclockwise turn image of the quadrilateral
   $ABCD$ about point $(0, 0)$. Do not use tracing paper.

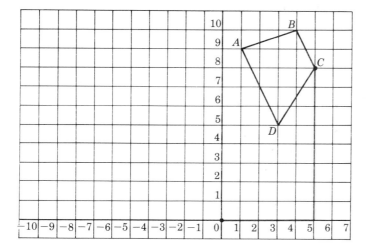

In the following exercises draw the figure on graph paper.

7. Draw in red the images of lines p and q under a 90° clockwise turn about point O.

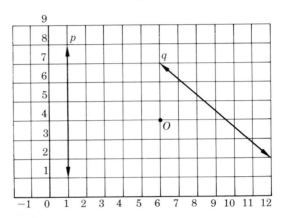

8. Draw in black a triangle and a line whose turn images, under the turn indicated, are the gray triangle and line.

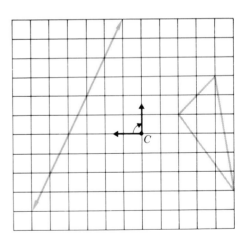

*9. Suppose lines p and q are perpendicular at A, and that q' is the image of q and A' the image of A under a turn.

Draw in red the image of line p under this turn. Try to locate the center of the turn.

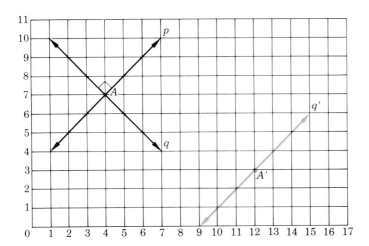

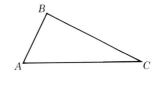

*10. Recall how to construct a line perpendicular to a given line at a given point on the line and then use a compass and straight edge to *construct* the 90° counterclockwise image of $\triangle ABC$ about point O.

*11. The fact that $(-5,3)$ is the 90° counterclockwise turn image of $(3,5)$ around the origin O is sometimes expressed as follows:

$$(3,5) \xrightarrow{O,90° \text{ ccl}} (-5,3)$$

Try examples with several specific points and complete the following:

a) $(x,y) \xrightarrow{O,90° \text{ ccl}} (?,?)$

b) $(x,y) \xrightarrow{O,90° \text{ cl}} (?,?)$

c) $(x,y) \xrightarrow{O,180°} (?,?)$

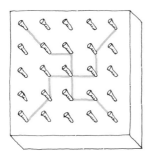

*12. If you place your finger on the center wall of this geoboard and turn the board 90° (clockwise or counterclockwise), the gray figure appears exactly as it is now. The figure is said to have 90° turn symmetry about the center of the geoboard.

How many different figures like this can you find on a 5 × 5 geoboard? Each figure you find should have the following characteristics.

a) It divides the geoboard into four sections that are the same size and shape.

b) The figure is formed by drawing four identical paths, starting at the center of the geoboard.

c) The paths are formed by drawing segments from one dot to another. Paths cannot cross. When a path reaches a point on the boundary of the geoboard array, it stops.

d) If a figure can be flipped or turned to look like another, the two figures should be considered the same.

There are 24 such figures. Can you find them all?

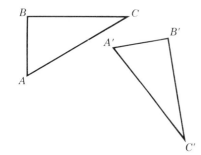

*13. The turn image of triangle ABC is triangle $A'B'C'$. Trace these triangles in this same position on tracing paper. Find the center of the turn. [*Hint*: Use the idea that the perpendicular bisector of a chord of a circle contains the center of the circle.]

*14. A 180° turn around a point is called a **half turn**.

$\triangle A'B'C'$ is the image of $\triangle ABC$ under a half turn. Trace these figures on your paper and devise a way to accurately find the point about which the half turn is made.

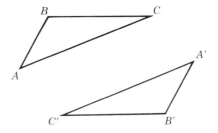

FLIPS

A flipping motion, just as the slide and the turn motion, is experienced in the real world. We flip the pages of a book or of a desk calendar. The admonition "flip it over" is appropriate for activity with a wide variety of objects, such as pancakes, cards, or records.

The main motivation for studying the motion of flipping, however, comes from its relationship to mirrors. This is so because the mirror seems to "flip" the object in front of it over and shows the result as the image in the mirror.

Architects and decorators have learned to use mirrors to achieve many interesting effects, and scientists have used them to build telescopes which probe deeply into outer space. Yet we take mirrors for granted, and many of us do not understand their basic properties. This can be illustrated by trying the experiments shown on the right.

Hinge two mirrors at a 90° angle and look into them as shown in this photo. How does your image differ from your usual image in a flat mirror?

Drawing by W. Miller, © 1962, New Yorker Magazine, Inc.

Hinge two mirrors at a 90° angle and look into them as shown. Are you surprised by what you see? Can you explain your observation?

Earlier we suggested that the three-mirror kaleidoscope provides exciting creative experiences for students. Other mirror problems such as those suggested on the preceding page and on page 41 of Chapter 2, can also broaden the students' horizons. Also students universally enjoy mirror writing, so don't forget the mirror as an interesting geometrical object to explore.

Many persons find it amazingly difficult even to visualize mentally the image of an object in a mirror. Mirror Cards* test one's ability to visualize and provide both a recreational and an educational experience for children of all ages.

Another handy instrument for experimenting with mirrors is a Mira, described in Chapter 2, p. 40. For example, place a Mira along a line in a plane which is perpendicular to the sheet of paper. Then any geometric figure on one side of the Mira has a "flip" image on the opposite side. By reaching around behind the Mira we can trace this image. (See Appendix A for activities which introduce the Mira.)

MIRA

Drawing by Chas. Addams, © 1957, New Yorker Magazine, Inc.

*These materials were developed for the Elementary Science Study by the Webster Division of McGraw-Hill Book Company.

This idea is illustrated in Fig. 5.7. Suppose that the Mira is placed along line *p*. Then the gray triangle would be the image of the black triangle.

If a cardboard cutout is placed to coincide with the black triangle, it needs to be "flipped" in order to make it coincide with the gray triangle. Hence, it is natural to refer to the gray triangle as either the "mirror image" or the "flip image" of the black triangle. The line *p* is called both the "mirror line" and the "flip line."

The stimulating drawing by M. C. Escher reproduced here utilizes a tessellation of the plane and a mirror to make the notion of a mirror image "come alive."

In a slightly less dramatic setting than that depicted by Escher, Investigation 5.3 will encourage you to use tessellations and mirrors to help develop your intuition about flips.

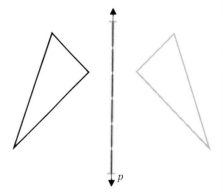

Figure 5.7

Magic Mirror, M. C. Escher

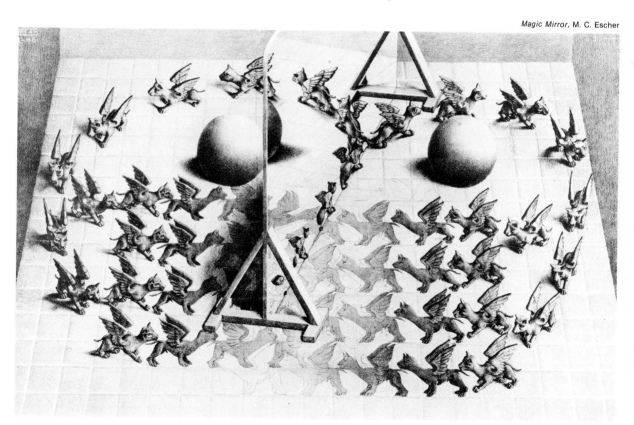

Investigation 5.3

A. What will you see if:

1. A piece of tinted plexiglass is placed along the flip line and you look at one side of it?

2. A mirror is placed along the flip line and you look in the mirror?

3. The "tessellation" is traced, folded together along the flip line, and you look through the folded paper at a bright light?

B. Trace the tessellation below. Can you show other mirror lines on the tessellation that are essentially different from the one above? (You may need to supply your own definition of the phrase "essentially different.")

C. How many essentially different mirror lines can you find for each of the following tessellations? (Show each line you find on a tracing of the tessellation.)

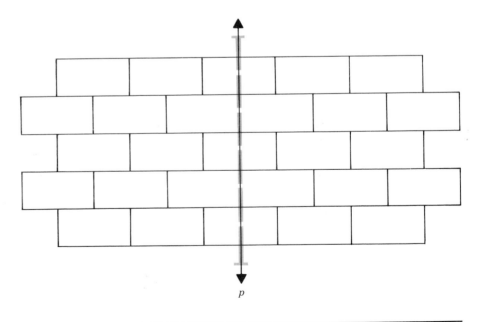

p

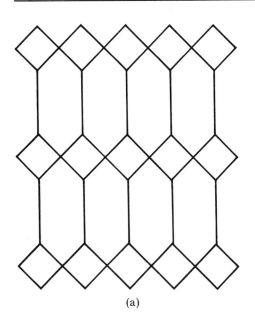

(a)

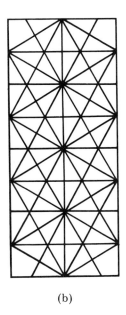

(b)

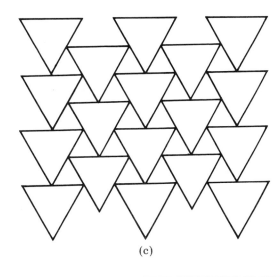

(c)

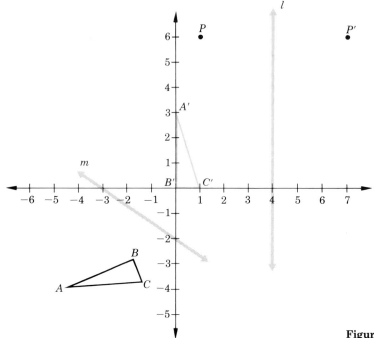

Figure 5.8

Since each flip involves a flip line, we usually describe a certain flip by indicating the flip line.

For example, in Fig. 5.8 point P' is the *flip image* of point P about flip line l, and P is the flip image of point P' about flip line l. Also, $\triangle A'B'C'$ is the flip image of $\triangle ABC$ about flip line m, and $\triangle ABC$ is the flip image of $\triangle A'B'C'$ about flip line m.

In Fig. 5.9 the gray figures are the flip images of the black figures (and visa versa) about the flip line. The reader should check this, using both a mirror and tracing paper.

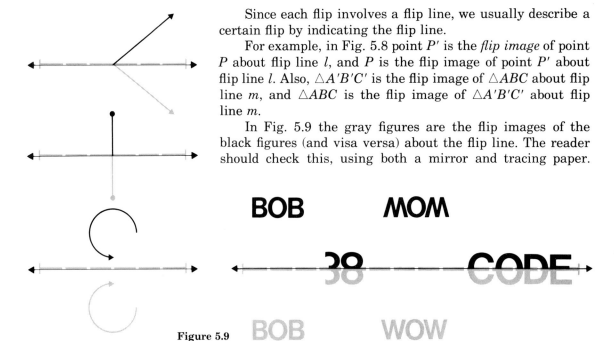

Figure 5.9

EXERCISE SET 5.3

1. The situations for this exercise are taken from the book *The Ambidextrous Universe* by Martin Gardner (New York: Basic Books, 1964).

   a) Place a Mira on the dashed line. How can you explain what you see?

   (a)

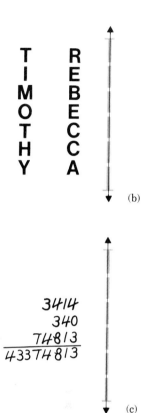

   (b)

   b) "A Mira is unbelievable, since it reverses girls' names, but doesn't reverse boy's names." Place your Mira on the dashed line. Now do you believe that this is true? Explain.

   c) Is this sum correct? Place your Mira along the dashed line to check it.

$$\begin{array}{r} 3414 \\ 340 \\ + \quad 74813 \\ \hline 43374813 \end{array}$$

   (c)

2. Draw $\triangle ABC$ and line l on your paper.
   Show how you can accurately draw the flip image of $\triangle ABC$ using (a) tracing paper, (b) tinted plexiglass, (c) paper folding, and (d) a ruler and compass construction.

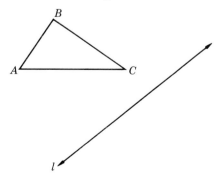

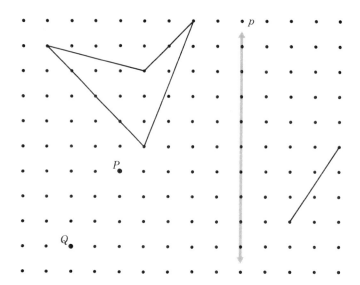

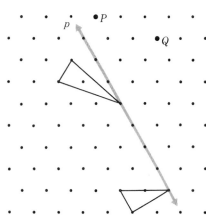

3. Copy these points and figures on *square dot paper*. Using only the dot paper (and a straight edge), draw the mirror image in line p of each of the figures shown above.

4. Copy these points and figures on *triangular dot paper*. Using only the dot paper (and a straight edge), draw the mirror image in line p of each of the figures shown at the left.

5. Draw the flip image in line p for each of the points in the figure below. Does this suggest that the flip image of an entire line is a line?

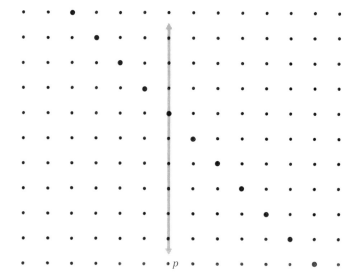

6. a) Draw a line q whose flip image in line p is parallel to p. How many such lines are there?

   b) Draw a line r whose flip image in line p is the line r again.

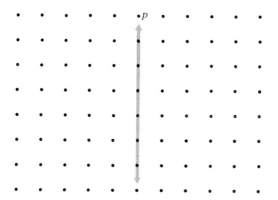

7. Use the figure at the right and draw lines p and q such that the flip image of A in p is B and the flip image of B in q is C.

8. a) Given lines p and q and the points A and B, draw the points A' and B', the mirror image of A and B in line p. Then draw the mirror image of A' and B' in line q and call them A'' and B''.

   b) How does distance AB compare with distance $A'B'$, and $A'B'$ with $A''B''$?

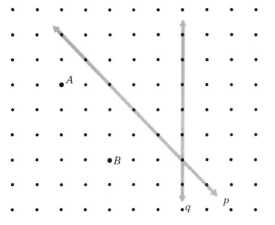

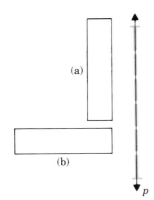

(a)

(b)

The reason this looks so strange is that it was written using a mirror.

A | A

B | B

↕ p

9. a) Which letters of the alphabet, like A, are mirror images in line p of themselves?

   b) Write several words which, when printed in rectangle (a) will have a mirror image in line p which is again a word.

10. Repeat Exercise 9 for rectangle (b).

11. Can you decode this message shown at the left?

12. If you place a mirror on line l, you form a word from the "half word" shown.

    a) How many such half words can you find?

    ← ▬▬ DOD ▬▬ → l

    b) How many different half numerals can you find?

13. Choose a tessellation, draw it, and analyze it with regard to reflectional symmetry.

COMBINING SLIDES, TURNS, AND FLIPS

It is only on rare occasions that we observe a physical situation involving motion in which a slide, a turn, or a flip occurs completely by itself. Instead, we often observe a myriad of combinations of these motions as, for instance, when a child on a sled slides down the hill, makes a sharp turn, and flips over. As in this example, we often miss the intricacy of the combination of motions because of a primary concern about the final result. In laboratory situations, as when a scientist investigates the motions of certain particles in a magnetic field, we would expect a much more careful analysis of the component motions.

    Investigation 5.4 provides an opportunity to consider the operation of combining motions, and to analyze the results of these combinations. Note that when two motions are combined, the image under the first motion becomes the starting figure for the second motion.

✳ The early study of combinations of slides, turns, and flips in a physical situation provides an interesting conceptual background for the later study of composition of functions. Many concepts develop slowly over a long period of time and it is desirable to begin the intuitive development of important concepts early.

Investigation 5.4

For each situation *A* through *C* draw a triangle on your paper and use tracing paper, compass and straight edge, or a Mira to perform first one motion on the triangle and then another.

In each case, can you *carefully* describe a single motion that has the same effect (produces the same final image) as the two motions in succession? Tell as much about the single motion as you can, and try to form a generalization about the situation. You may need to construct more examples of your own to formalize these generalizations.

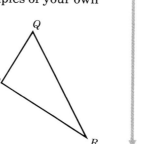

A. First: Find the flip image in line *r*.
    Then: Find the flip image in line *s*.

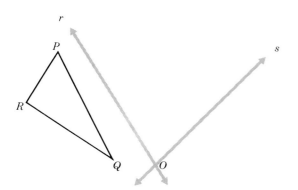

B. First: Find the flip image in line *r*.
    Then: Find the flip image in line *s*.

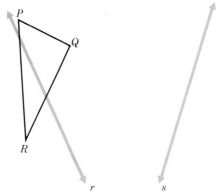

C. First: Find the flip image in line *r*.
    Then: Find the flip image in line *s*.

When two motions are combined, as in Investigation 5.4, we often call the single motion that has the same effect **the product of the two motions**. Just as the product of two whole numbers is another whole number, the product of two motions is another motion. *Which of these conjectures do you think are true about the product of two motions? Can you change those that are false in such a way as to make them true?*

Conjecture 1. *The product of two flips in parallel lines produces the same effect as a single flip.*

Conjecture 2. *The product of two flips in parallel lines produces the same effect as a slide in a direction perpendicular to the lines and through a distance equal to the distance between the lines.*

Conjecture 3. *The product of two flips in intersecting lines produces the same effect as a turn about the point of intersection through an angle equal to the angle between the two lines.*

After a careful analysis of both your investigation and the above three conjectures, you should decide that all three conjectures are false. Figure 5.10 shows that if lines r and s are parallel, the two flips have the same effect as a slide with direction perpendicular to the lines r and s and through a distance $2d$, with d the distance between the lines.

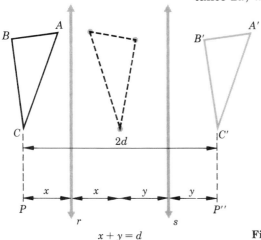

$$x + y = d$$

Figure 5.10

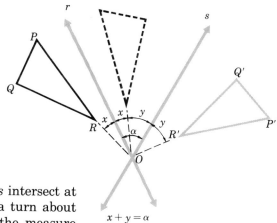

$$x + y = \alpha$$

Figure 5.11

Figure 5.11 convinces us that if lines r and s intersect at point O, the two flips have the same effect as a turn about point O and through an angle 2α where α is the measure between the two lines. To summarize, we see that the result of two flips is either a slide or a turn, depending upon whether the two flip lines are parallel or intersecting.

So far in this section we have considered only the possible single motions resulting from the combination of two flips. A detailed analysis of all possible combinations—such as the result of combining two turns with different centers, the result of combining a slide and a turn, etc.—is beyond the scope of this chapter. A few of these possibilities will be considered in the exercises and in Chapter 6.

It should be mentioned, however, that the motion resulting from the combination of a slide and a flip is usually not itself a slide, a turn, nor a flip. This motion, described and exemplified in Fig. 5.12, will be called a **slide-flip** for the purposes of this chapter.

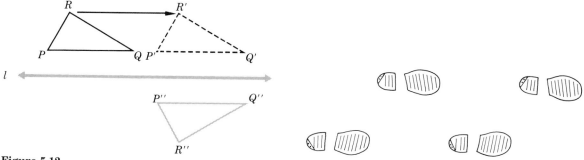

Figure 5.12

A slide RR' followed by a flip in line l ($l \| RR'$) results in a motion we call a "slide-flip" which is different from a slide, a flip, or a turn.

The pattern made by a person's footprints in the snow is an example of a sequence of "slide-flips."

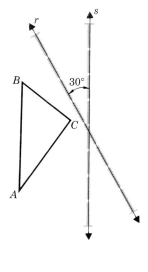

EXERCISE SET 5.4

1. Draw parallel lines p and q on your paper 4 cm apart.

   a) Use the information in this section to describe the slide that is the result of a flip in line p followed by a flip in line q. Give a careful description of both the distance and the direction of the slide.

   b) Use tracing paper or a Mira to carry out the flips and check to see that the description you gave in (a) is correct.

2. Draw intersecting lines r and s on your paper with angle of intersection 30°.

   a) Use the information in this section to describe a turn that is the result of a flip in line r followed by a flip in line s. Give a careful description of the center, angle, and direction of the turn.

   b) Use tracing paper or a Mira to carry out the flips and check to see that the description you gave in (a) is correct.

3. Draw lines t and u that are perpendicular to each other. Describe the result of a flip in line t followed by a flip in line u. Carry out the flips to see if your description is correct.

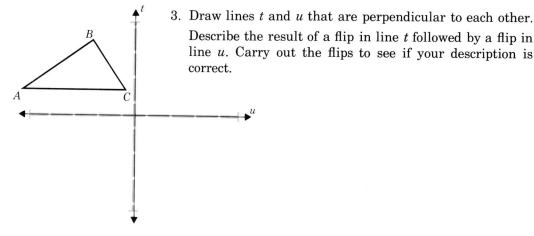

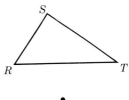

4. A 180° turn about a point is called a **half turn**. Mark points X and Y on your paper and draw $\triangle RST$.

    a) First complete a half turn around X. Using the image of this turn, complete a half turn around Y.

    b) What single motion has the same effect as a half turn about X followed by a half turn about Y? Be as specific as possible in describing this motion.

5. $\triangle A'B'C'$ is the image of $\triangle ABC$ under a slide.

    Trace these figures. Can you construct a second line n so that a flip about the line m followed by a flip about the line n will have the same effect as the slide?

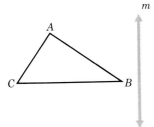
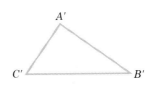

6. $\triangle A'B'C'$ is the image of $\triangle ABC$ under a turn about center Q (test this with tracing paper).

    Trace these figures. Can you construct line q so that a flip about the line p followed by a flip about the line q will have the same effect as the turn?

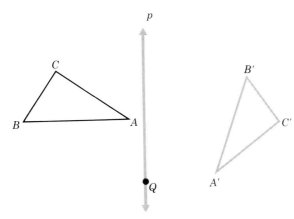

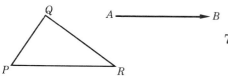

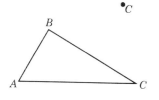

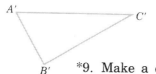

7. Trace these figures on your paper. Find the half-turn image of △PQR about point C followed by the slide AB.

   What single motion has the same effect as this sequence of motions? Describe this single motion as carefully as possible.

8. Triangle A'B'C' is the image resulting from first performing a slide and then performing a flip with △ABC. We say that △A'B'C' is the *slide-flip* image of △ABC.

   The flip line can be found by connecting the midpoints of AA' and BB'.

   Trace the two triangles shown here on your paper and find the flip line. Use tracing paper to see if the line you found is correct.

*9. Make a cardboard rectangle with a gray arrow on one side and a black arrow on the opposite side. Both arrows should point in the same direction.

Suppose you have the cardboard rectangle lying inside a rectangular outline on a piece of paper. Here are descriptions of possible motions that you can make with the rectangle so it will fit back inside the outline again.

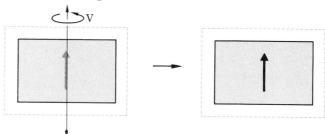

Flip the rectangle about a vertical line of symmetry. This is a V flip.

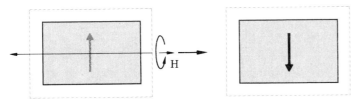

Flip the rectangle about a horizontal line of symmetry. This is an H flip.

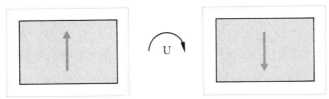

Turn the rectangle 180° about the center (clockwise or counterclockwise). This is the U (upside down) turn.

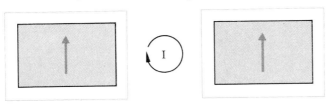

Turn the rectangle 360° (or leave it unchanged). This is the I (identity) turn.

a) Complete this "product table" showing all possible products of pairs of these motions.

b) What patterns do you see in the table?

Followed by	I	U	H	V
I				
U				
H				U
V				

*10. Make a 3-inch cardboard square with a black arrow on one side and a gray arrow on the opposite side. Place the square inside a square outline.

a) Describe all possible motions of the square which would allow it to fit back inside the outline again.

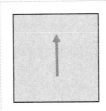

b) Make a product table showing all possible products of pairs of these motions.

c) What patterns do you see in the table?

d) We say that the set {I,R,L} is **closed** with respect to the operation of combining motions because the product of any two of the motions in the set is another motion in the set. The product table shows this to be true.

How many more sets of motions of the square can you find that are closed with respect to the operation of combining motions? Show the product table for each.

	I	R	L
I	I	R	L
R	R	L	I
L	L	I	R

R: 90° clockwise turn
L: 90° counterclockwise turn

*11. Investigate the motions of an equilateral triangle and make a product table.

Describe any interesting things you discover about this situation.

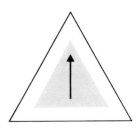

*12. Explore the product of three half turns. What generalizations can you make about the product of an odd number of half turns? an even number? Draw pictures to support your conclusions.

*13. Recall that the operation of multiplication of whole numbers is commutative. That is, $a \cdot b = b \cdot a$. Can you combine the two motions indicated in each part below in either order and get the same result? (always? sometimes? never?) Draw pictures to support your conclusions.

a) two flips

b) two sides

c) two half turns

d) two turns about point C

*14. Experiment with tracing paper and the Mira to complete one unshaded entry in the table by describing the resulting motion(s) in as much detail as possible.

Followed by	Slide	Turn	Flip
Slide			
Turn			
Flip			

MOTIONS AS A COMBINATION OF FLIPS

In the last section we began with two mirror lines and explored the product of flips. In this next investigation we explore the converse situation. We begin with a pair of figures and seek a mirror line, or several mirror lines, such that the product of the flips sends one figure onto another.

Investigation 5.5

For each situation below construct a line p such that the flip image of the "black figure" in p is the "gray figure". If you cannot find a single line p, search for two lines p and q such that the gray figure is the product of the flip in p and q. If two lines are not sufficient, use three lines.

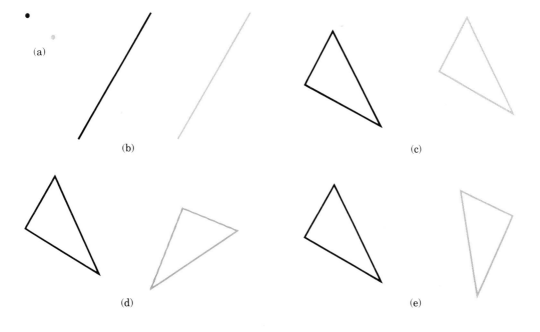

As you completed Investigation 5.5 you undoubtedly noticed that certain parts were easier to complete than others.

Part (a) simply required the construction of the perpendicular bisector of the segment determined by the gray and black points. The theme of "perpendicular" can be used to complete parts (c), (d), and (e). The following figures illustrate a solution to part (e).

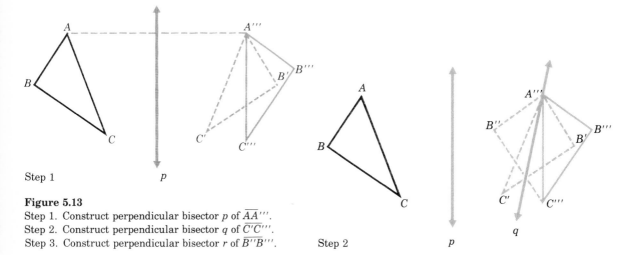

Figure 5.13
Step 1. Construct perpendicular bisector p of $\overline{AA'''}$.
Step 2. Construct perpendicular bisector q of $\overline{C'C'''}$.
Step 3. Construct perpendicular bisector r of $\overline{B''B'''}$.

Investigation 5.5 and Fig. 5.13 suggest the following generalizations..

1. Given a geometric figure (e.g., a triangle) and its image under a *slide*: A pair of parallel lines p and q can be found so that the image under flip in lines p and q takes the original figure onto the slide image of the figures.

2. Given a geometric figure and its image under a *turn*: A pair of lines p,q intersecting in the center of the turn can be found so that the image under flip in lines p and q takes the original figure onto the turn image.

3. Given a geometric figure and its image under a *slide-flip:* A set of three lines p, q, r exist such that the image under flip p, q, and r takes the original figure to the final position under the given motion.

These generalizations will be accepted as true without proof and may be used in the following exercises.

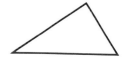

EXERCISE SET 5.5

1. Given a triangle and its image under a slide:

   a) Draw a pair of lines so that the product of flips in the two lines takes the black triangle onto the gray triangle.

   b) How many different pairs of lines such as those in part (a) can you find?

2. Draw a triangle and use tracing paper to find its image after a turn.

   a) Draw a pair of lines so that the product of flips in the two lines takes the original triangle onto its image.

   b) How many different pairs of lines like those in part (a) can you find?

3. If p and q are parallel, how does the final position of △ABC, after flipping through p, then q, compare to the final position after flipping through q, then p? Is the result ever the same?

4. If p and q intersect, how does the final position of △ABC, after flipping through p, then q, compare to the final position after flipping through q, then p? Is the result ever the same?

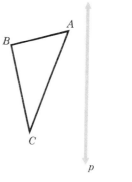

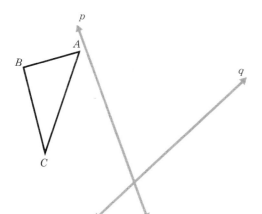

5. Which of the following conjectures do you think are correct (i.e., have been proven to be true)?

   a) If $\triangle ABC$ and $\triangle A'B'C'$ are the same size and shape, then one of these triangles can be made to coincide with the other using the product of two flips. [*Note*: The two triangles may not have the same orientation.]

   b) Any slide or turn can be described as the product of two flips.

   c) If $\triangle ABC$ and $\triangle A'B'C'$ are oppositely oriented, then the two triangles can never be made to coincide after two flips.

BIBLIOGRAPHIC REFERENCES

[2], [3], [17, 25-29], [30], [39, 1-7, 25-32], [41, 162-170], [43, 114-121], [63], [102]

PEDAGOGICAL ACTIVITIES FOR THE TEACHER

1. Read the article, "Geometry All Around Us—K–12" by John Egsgard in the October 1969 issue of *The Arithmetic Teacher*. In the chart which he gives at the end of the article, study the role of transformations and motions in geometry and react to his suggestions relative to grade levels of introduction of topics.

2. Study the books entitled "*Motion Geometry*" by Phillips and Zwoyer, published by Harper and Row, Publishers. Decide at what grade level this material might be appropriate and give specific reasons for your decision.

3. At least one basic text series for the elementary school presents ideas of slides, flips, and turns in the middle grades. Find such a series and develop an outline which specifies what type of material it presents and at what grade levels. Give page numbers of parts of the textbooks where these ideas are developed. Make any additional suggestions as a result of your study of this chapter that you think might improve this development for elementary school children.

4. Consult a reference such as pages 233–295 of the Thirty-fourth Yearbook of the National Council of Teachers of Mathematics, *Instructional Aids in Mathematics,* and

    a) devise a teaching aid which would enable middle-grade children to easily find the rotation image of a triangle through any number of degrees; and

    b) devise teaching aids which would enable students to perform slides and flips mechanically.

5. Plan a lesson, including all appropriate teaching aids, which could be used to help junior high school or high school students understand how to combine motions and to form generalizations about these combinations. Try your ideas with an individual student or small group of students and report your results. Suggest specific ways you would improve the activity for use in your classroom.

6. Plan an 8 mm film loop which would graphically demonstrate the meaning of slides, turns, and flips.

7. Make a list of questions about mirrors which you could ask students at a given level. You might want to refer to a book such as *The Ambidextrous Universe* by Martin Gardner for ideas. Ask a student or group of students your questions. Provide mirrors so that they might experiment and answer the questions. Report on their knowledge of mirrors and their reactions to your questions.

6

Translations, Rotations, Reflections, and Congruence

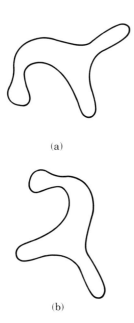

(a)

(b)

Figure 6.1

INTRODUCTION

In Chapter 5 we studied slides, turns, and flips as examples of motions in the physical world. As a result of your study you have probably developed several ideas about motions—some well understood and some understood barely above the subconscious level. A simple observation which you undoubtably made is that a geometric figure maintains its "size and shape" as it slides, turns, or flips. This observation provides motivation for a once popular definition of "congruence." Prior to the more careful consideration of the foundations of school mathematics which has occurred during the past decade or so, a common definition stated that *two figures are congruent if they can be made to coincide.* According to this definition the two figures in Fig. 6.1 are congruent. For if Fig. 6.1 (a) is traced onto a piece of acetate, this tracing can be slid, flipped, and turned to coincide with Fig. 6.1 (b). While this description of congruence is adequate when on the intuitive, physical level, it is inadequate in a more careful mathematical development for at least two reasons. First, it is not possible to actually "pick up" one set of points and move them until they "coincide" with another set. Second, it is not clear, even on the intuitive level, what it means to make a solid object coincide with another solid object. For example, examine two blocks, both $1'' \times 2'' \times 4''$. It is not possible to move one so that it coincides with the other since one block cannot pass through the other. The model of tracing on acetate which we used in discussing Fig. 6.1 does not apply here either. Yet, these two blocks have the same size and shape, and consequently our definition of congruence should be broad enough to include these two blocks as a pair of figures that are congruent to one another.

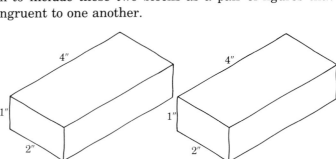

We can remedy the deficiences of the early definition, maintain the same intuitive flavor, and at the same time present a definition of congruence broad enough to encompass all plane figures and solids by employing the mathematical abstractions of slides, turns, and flips. These abstractions, known as translations, rotations, and reflections, are the subject of the first few pages of this chapter.

FROM SLIDES TO TRANSLATIONS

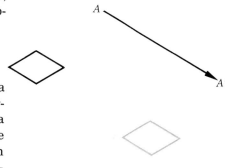

We have seen in Chapter 5 that when an object is slid across a plane surface, the final position of the object can be determined from its initial position by specifying a direction and a distance. For example, in the figure, a parallelogram shape can be placed on the black outline and made to coincide with the gray outline by sliding it through the distance and direction specified by the arrow AA'.

A segment together with a direction is called a **directed segment** and is denoted $\overrightarrow{AA}'$. Two directed segments $\overrightarrow{AA}'$ and $\overrightarrow{BB}'$ of the same length and direction determine the same slide; hence, we shall call directed segments $\overrightarrow{AA}'$ and $\overrightarrow{BB}'$ **equivalent directed segments** if they are of the same length and direction.

Now suppose that instead of only the parallelogram sliding, we imagine that the entire plane slides. We can visualize this physically as follows. Place a piece of transparent acetate over a piece of paper on which several points A, B, and C have been drawn and trace these points onto the acetate, labeling them A', B', and C'. When the acetate is slid, the points A', B', and C' represent the images of points A, B, and C after the slide AA'. However, it is evident that not only points A, B, and C possess images after slide AA', but each point in the plane has a unique image point after this slide. That is, the physical motion of sliding the acetate has established a correspondence which associates each point P of the plane with a new point P' called its slide image.

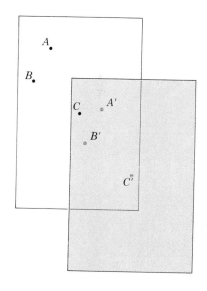

The interesting question which arises is, *Can this correspondence between points be described without resorting to the physical model of sliding acetate?* The answer is yes. The particular definition given below has evolved over a period of years as an appropriate way to describe this correspondence.

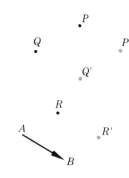

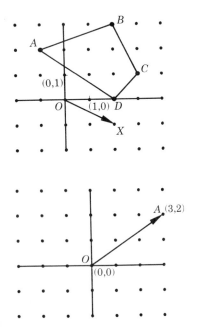

The concept of function (as a unique type of correspondence) is very important in mathematics. The intuitive work with slides, flips, and turns in the elementary school provides a valuable background for an understanding of the idea of transformation, which is a geometric type of function. Thus early physical experiences provide the foundation for an important mathematical idea.

Definition. *Suppose A and B are a pair of points in the plane and consider the directed segment $\vec{AB}$. Each point P can be associated with the unique point P' where the directed segment $\vec{PP'}$ has the same direction and length as the given directed segment $\vec{AB}$. This correspondence between points is called the* **translation associated with the direct segment** $\vec{AB}$ *and is denoted S_{AB}.*

The point P' associated with P is called the **image of P under the translation** S_{AB}. We write $S_{AB}(P) = P'$.

Note that it is not the physical motion which we call the translation—it is the *correspondence* between points established by the directed segment which we are calling the translation. So a slide is a physical motion and a translation is a correspondence between points—the mathematical abstraction of the physical motion. A careful consideration of the physical world has led to a mathematical abstraction. This procedure of transforming observations from the physical world into an abstract description occurs frequently in mathematics. It is the abstract descriptions which mathematicians study.

EXERCISE SET 6.1

1. Suppose the plane has the rectangular coordinate system as indicated in the figure at the left.

   a) What are the coordinates of the vertices of the quadrilateral $ABCD$?

   b) What are the coordinates of the vertices of the quadrilateral $A'B'C'D'$ where $A' = S_{OX}(A)$, $B' = S_{OX}(B)$, $C' = S_{OX}(C)$, and $D' = S_{OX}(D)$?

2. Consider the translation S_{OA} where O and A have coordinates $(0, 0)$ and $(3, 2)$, respectively, as shown at the left.

   a) For each point below find the coordinates of its image under the translation S_{OA}.
      i) $(-1, 1)$ ii) $(4, -2)$ iii) $(7, 5)$ iv) (x, y)

   b) i) Find the point whose image under S_{OA} is $(1, 3)$ (We call this point the pre-image of $(1, 3)$.)

      ii) Find the point whose image under S_{OA} is $(-2, -6)$.

      iii) Find the point whose image under S_{OA} is (x, y).

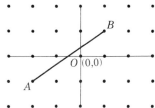

(These questions could be rephrased as, find the pre-image of $(1,3)$, $(-2,-1)$ and (x,y) with respect to the translation S_{OA}.)

3. Find the coordinates of the point C such that the translations S_{AB} and S_{OC} are the same translation. (You need to think carefully about what a translation actually is — a correspondence between points — in order to decide what must be meant by "same translation.")

4. a) Describe the location of a point B such that the image of line s under S_{AB} is line t.

   b) How many choices were there for point B?

*5. Given points A, B, C, D, and P. Draw

   a) $S_{AB}(P)$ and $S_{BA}(P)$

   b) $S_{CD}\,[S_{AB}(P)]$

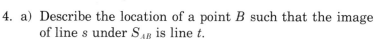

*6. Given lines s and t and segment $\overline{AB}$ as indicated. Describe how to use a suitable translation to construct a segment $\overrightarrow{PQ}$ such that $P \in s$, $Q \in t$, and $\overrightarrow{PQ}$ and $\overrightarrow{AB}$ are equivalent.

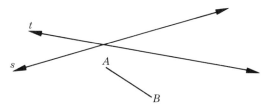

FROM TURNS TO ROTATIONS

In our analysis of tessellations in earlier chapters we observed that for the regular tessellation of equilateral triangles printed in Fig. 6.2, a triangular cutout which is placed to coincide with a shaded triangle cannot be made to coincide with a white triangle by sliding only; a turning motion is also required. For example, region A can be turned clockwise about point X through $60°$ to coincide with region B, or through $180°$ to coincide with region C. Likewise region A can be turned about the point Y through $180°$ to coincide with region D.

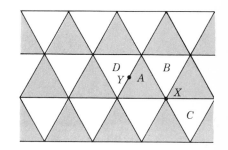

Figure 6.2

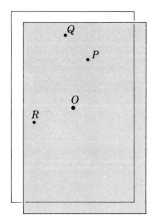

Now suppose we think of turning about a point not only a triangular region but the entire plane. We can visualize this physically by placing a piece of transparent acetate over a piece of paper and placing a paper fastener through them. Suppose we begin by labeling several points P, Q, and R on the paper and then tracing them onto the acetate. By turning the acetate sheet about the point O (the paper fastener), each of the points P, Q, and R move to new points P', Q', R' called the images of points P, Q, and R after the turn. When the acetate is turned counterclockwise through an angle of 35°, we notice that points P', Q', and R are positioned so that $\angle POP'$, $\angle QOQ'$ and $\angle ROR'$ all have the measure of 35°. It is clear that not only P, Q, and R possess images under this turn, but each point X of the plane has an image X' where the measure of $\angle XOX'$ is 35°. As in the case of the slide, the physical motion of turning the acetate has established a correspondence which associates each point P of the plane with a new point P' called its turn image.

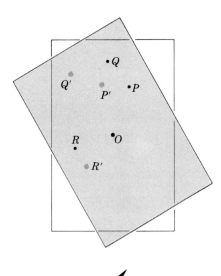

This correspondence, called a **rotation**, can be described without resorting to the acetate sheet. However, before defining what we shall call the *rotation through angle α with center O*, we digress briefly to describe a more general notion of angle measure which permits the possibility for angles with negative measure and eliminates the need to refer to clockwise or counterclockwise rotations.

Given an angle, say $\angle AOB$, we can designate one ray, say $\overrightarrow{OA}$, as the initial side and the other ray, $\overrightarrow{OB}$, as the terminal side. When this is done we say we have a *directed angle with initial side $\overrightarrow{OA}$*, and we denote this $\angle AOB$. Consequently, $\angle AOB$ and $\angle BOA$ are the same angles; but the directed angles $\angle AOB$ and $\angle BOA$ are not the same since the initial side and terminal side have been reversed. These angles are sometimes represented pictorially as in Fig. 6.3.

Figure 6.3

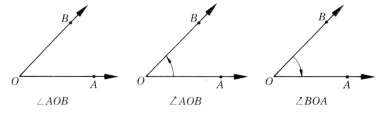

$\angle AOB$ $\angle AOB$ $\angle BOA$

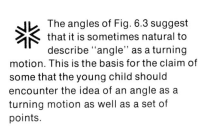

The angles of Fig. 6.3 suggest that it is sometimes natural to describe "angle" as a turning motion. This is the basis for the claim of some that the young child should encounter the idea of an angle as a turning motion as well as a set of points.

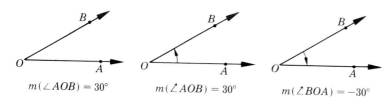

$m(\angle AOB) = 30°$ $m(\angle AOB) = 30°$ $m(\angle BOA) = -30°$

Figure 6.4

Before assigning a measure to directed angles we need to designate one direction as the positive direction and the other direction as the negative direction. We shall agree that $m(\angle AOB)$ is positive if the cyclic orientation of the triple (OAB) is counterclockwise, and that $m(\angle AOB)$ is negative if the cyclic orientation of the triple (OAB) is clockwise. Generalizing this scheme to all directed angles, we see in Fig. 6.4 that a directed angle can be associated with a degree measure between $-180°$ and $+180°$. We are ready for the definition.

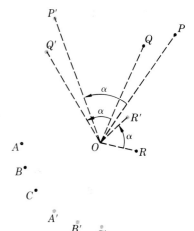

Definition. *Consider a point O and a particular angle measure α, $-180° < \alpha \le 180°$. Each point P, distinct from O, can be associated with the unique point P' in the plane where $m(\angle POP') = \alpha$ and distance OP is equal to distance OP'. This correspondence which associates the point O with itself and each point P, distinct from O, with the point P' is called the* **rotation with center** *O* **and angle** *α and is denoted $R_{O,\alpha}$. We write $R_{O,\alpha}(P) = P'$.*

EXERCISE SET 6.2

1. Use a protractor to measure on the figure at the right each of the directed angles.

   a) $m(\angle BOC) =$ b) $m(\angle AOB) =$
   c) $m(\angle COA) =$ d) $m(\angle COB) =$

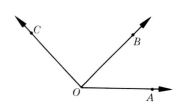

2. Trace these points and use a protractor and compass to draw the images indicated below.

   a) $R_{O,30°}(P)$ b) $R_{O,45°}(Q)$
   c) $R_{O,-60°}(Q)$ d) $R_{O,75°}(Q)$
   e) $R_{O,60°}(P), R_{O,60°}(Q),$
      $R_{O,60°}(S),$ and $R_{O,60°}(R).$

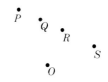

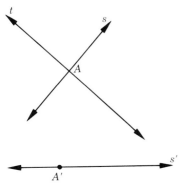

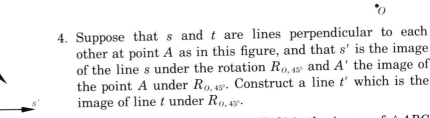

3. a) Trace the points P, Q, and O and draw points A, B such that $R_{O,60°}(A) = P$ and $R_{O,60°}(B) = Q$.

   b) For what angle α is the following true?

   $$R_{O,\alpha}(P) = A;\ R_{O,\alpha}(Q) = B.$$

4. Suppose that s and t are lines perpendicular to each other at point A as in this figure, and that s' is the image of the line s under the rotation $R_{O,45°}$ and A' the image of the point A under $R_{O,45°}$. Construct a line t' which is the image of line t under $R_{O,45°}$.

5. a) In the diagram below $\triangle A'B'C'$ is the image of $\triangle ABC$ under a rotation through $90°$ centered at the origin. What are the coordinates of A', B', C'?

   b) Generalize the results from part (a). The image of the point (x,y) is the point $(\underline{}, \underline{})$.

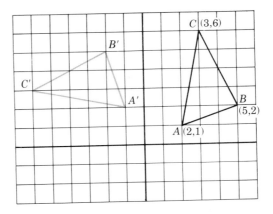

FROM FLIPS TO REFLECTIONS

We introduced the notion of line reflection in Chapter 5 through a discussion of mirrors and the flipping motion. In this section we shall build upon these experiences and describe the concept of "line reflection." As we develop this subject, particularly when we include three-dimensional considerations, it will become evident that this concept corresponds more closely to the model of the "mirror" than to the model of "flipping." So we shall build our discussion in this chapter on the mirror.

We know from our experience with mirrors or the Mira that if a line l is drawn on paper with a straight edge, and if a Mira is placed on this line, then each point P in front of the mirror has an image which appears to lie behind the mirror on a line perpendicular to l and the same distance from l as the point P. We say that the point P is reflected in line l with image P'. Also each point R behind the plexiglass appears (when viewed from behind) to have an image R' in front of the plexiglass. Thus each point of the plane can be associated with the aid of a mirror with another particular point in the plane.

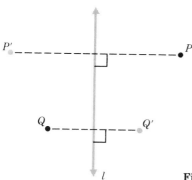

Figure 6.5

Moving from the physical to the abstract, imagine that the line itself acts as a mirror, as shown in Fig. 6.5. We can describe the image points P and Q not on line l in terms of the relations of perpendicularity and midpoint. The definition of line reflection is as follows.

Definition.   *Let l be a line. Each point P not on line l can be associated with the unique point P' where l is the perpendicular bisector of $\overline{PP'}$. The correspondence which associates each point P on l with the point itself and each point P not on l with the point P' is called the* **line reflection in line** l *and is denoted M_l. If P is any point, we write $M_l(P) = P'$.*

EXERCISE SET 6.3

1. a) Given $\triangle ABC$ below, construct the image $\triangle A'B'C'$ of $\triangle ABC$ under the line reflection M_x. Construct the image of $\triangle A''B''C''$ of $\triangle ABC$ under the line reflection M_y.

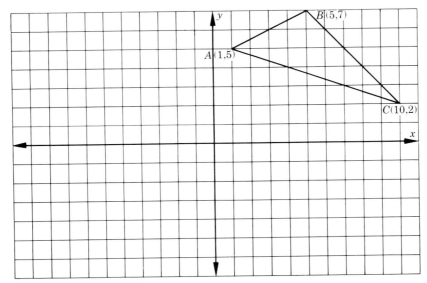

b) What are the coordinates of the points?

$M_x(A) =$	$M_x(B) =$	$M_x(C) =$
$M_y(A) =$	$M_y(B) =$	$M_y(C) =$

c) If P is the point with coordinates (x,y), what are the coordinates of $M_x(P)$ and $M_y(P)$?

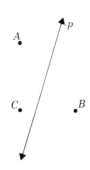

2. Use this figure and draw points A', B', and C' such that $M_p(A) = A'$, $M_p(B) = B'$, and $M_p(C) = C'$.

3. Consider the line reflection M_l defined by $M_l(x,y) = (y,x)$. Discover by drawing figures and their images which line in the plane is the reflection line.

4. Given lines s and t as shown at the right:

a) Select any pairs of points P and Q on line t.

b) Use a compass to construct the image points P' and Q' under M_s.

c) Draw the image of line t under M_s.

5. A point P is a **fixed point** for a line reflection M_p if $M_p(P)$ $= P$. Describe the fixed points for a line reflection M_p.

6. a) Construct a line t such that $M_s(t)$ is parallel to t.

b) Can you describe *all* lines t such that $M_s(t) \| t$?

c) Describe all lines t such that $M_s(t) = t$.

7. Given lines s and t as shown at the right: Draw a line u such that its image under M_s is parallel to line t.

8. Given a line t:

a) Construct a square whose image under M_t is the square itself.

b) Construct a pentagon whose image under M_t is the pentagon itself.

c) Construct a triangle with no vertex on line t whose image under M_t is the triangle itself.

d) Construct a parallelogram (not a rectangle) whose image under M_t is the parallelogram itself.

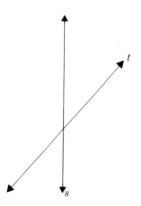

*9. Suppose A is the position of a cue ball on a pool table. Find the point on each of the four cushions which the cue ball should strike (disregarding spin) in order to rebound and strike ball B squarely. [*Hint*: Reflect point A in each of the four sides of the table.]

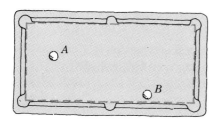

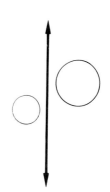

*10. Construct an equilateral triangle with one vertex on each of the two circles, and the other vertex on the line t.

ISOMETRIES

Euclidean geometry is now being taught from the point of view of transformations in many secondary school geometry courses. Intuitive experiences with motions from kindergarten on up provide a valuable foundation for this work. Teachers interested in such a secondary level course should consider the text by Coxford and Usiskin entitled *Geometry – A Transformation Approach* (Laidlow, 1971).

Translations, rotations, and line reflections are each a correspondence between points which has been motivated by a physical situation. For each of these correspondences each point of the plane is associated with a unique point of the plane. A correspondence of this type is called a transformation. Summarizing, we state:

A transformation from the plane to the plane is a correspondence which associates each point of the plane with one and only one point of the plane. We write $T(P) = P'$ to indicate that the point P corresponds to point P' under transformation T. We say that P' is the image of P under T.

Whenever a rectangular coordinate system is assigned to the plane, transformations can be described algebraically. For example, consider the transformation T defined by $T(x,y) = (2x, 3y + 1)$. The image of point $(3, 2)$ can be easily computed by $T(3, 2) = (2 \cdot 3, 3 \cdot 2 + 1) = (6, 7)$. In Investigation 6.1 we shall explore transformations which are defined algebraically like the above example.

Investigation 6.1

For each of the transformations defined below the image of a triangle is a triangle. (This property is not true for all transformations.)

A. For each transformation T below:

1. Sketch on graph paper $\triangle PQR$ and its image under T, where $P = (1, 1)$, $Q = (1, 2)$, $R = (3, 1)$.

2. Decide whether the transformation is a translation, rotation, reflection, or none of these three.

a) $T(x,y) = (x + 3, y - 2)$ b) $T(x,y) = (-y, x)$
c) $T(x,y) = (y, x)$ d) $T(x,y) = (-x+2, -y+4)$
e) $T(x,y) = (-2x, \frac{1}{2}y)$ f) $T(x,y) = (x + 3, -y)$

B. In the above cases search for centers of rotation and reflection lines whenever they exist.

In Investigation 6.1 you discovered that there are transformations which are not translations, rotations, or reflections. For example, we see that the transformation $T(x,y) = (-2x, \frac{1}{2}y)$ maps the black triangle onto the gray triangle in Fig. 6.6. It is evident that this T does not preserve the size and shape of triangles and hence cannot be a translation, rotation, or reflection.

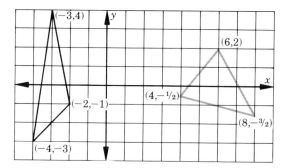

Figure 6.6

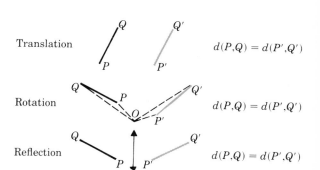

Figure 6.7

Since "size and shape" are such important geometric concepts we are particularly interested in transformations which do preserve "size and shape." "Size and shape" preserving transformations are precisely the transformations which preserve "distance." Such transformations are known as isometries (iso-equal, -metry-distance). The diagram in Fig. 6.7 makes plausible the claim that translations, rotations, and reflections are isometries.

The formal definition of isometry is the following one.

Definition. *A transformation T is an* **isometry** *if for each pair of points P and Q, the distance between P and Q is equal to the distance between P' and Q' ($P' = T(P)$ and $Q' = T(Q)$). That is, T is an isometry if $d(P,Q) = d(T(P), T(Q))$ for each pair of points P and Q.*

When a rectangular coordinate system is assigned to the plane, the distance between points $P(x_1, y_1)$ and $Q(x_2, y_2)$ is given by the distance formula

$$d(P,Q) = \sqrt{(x_1 - x_2)^2 + (y_1 - y_2)^2}.$$

The beginning ideas of coordinate geometry are often introduced in second or third grade. Piaget indicates in the book *The Child's Conception of Geometry* (Harper & Row, 1960) that the child makes definite progress around ages eight and nine in applying and even constructing systems of horizontal and vertical reference. Thus, the concept of rectangular coordinates can appropriately be initiated in the elementary school.

This formula can often be used to verify that a particular transformation is an isometry. For example, the reflection M_y sends point (x_1, y_1) to point $(-x_1, y_1)$ and (x_2, y_2) to point $(-x_2, y_2)$. The distance formula shows that the distance between any two points P, Q is equal to the distance between points P' and Q':

$$\sqrt{(-x_1 - (-x_2))^2 + (y_1 - y_2)^2} = \sqrt{(x_1 - x_2)^2 + (y_1 - y_2)^2}$$

$$d(P', Q') = d(P, Q).$$

Since the distance between two points is equal to the distance between their images, we say that distance is preserved by isometries. There are many other geometric properties which are preserved by isometries. We list and illustrate a few of these properties.

Betweenness is preserved. If B is between A and C, then B' is between A' and C'.

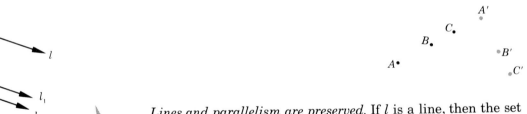

Lines and parallelism are preserved. If l is a line, then the set l' of all images of points of l is also a line. If l_1 and l_2 are parallel, then l'_1 and l'_2 are parallel.

Rays and segments are preserved. The image of a ray AB is a ray $A'B'$, and the image of a segment CD is a segment $C'D'$.

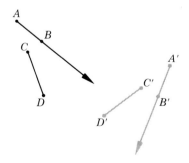

Angles and their measure are preserved. Since the image of a ray is a ray, it follows that the image of an angle is an angle. The measure of an angle and its image are equal.

Polygons and their size and shape are preserved. The image of a polygon is a polygon of the same shape.

Discovering properties which are preserved by isometries and other transformations is an important part of the study of geometry. Felix Klein in 1871 was the first to enunciate the point of view that geometry is the study of those properties which are preserved by certain types of transformations. This point of view motivates the interest in transformations today.

CONGRUENCE OF PLANE FIGURES

The concept of isometry provides an effective way of describing the relation of "congruence" often referred to more informally as the relation of "same size and shape." The many experiences we have had with motions in Chapter 5 suggest the following definition.

Definition. *Two geometric figures in the plane (i.e. subsets of the plane) $\mathscr{F}$ and $\mathscr{F}'$ are **congruent** if there exists an isometry T such that $T(\mathscr{F}) = \mathscr{F}'$.*

This definition of congruence is an extremely broad one which applies not only to triangles, quadrilaterals, and arbitrary polygons, but to all geometric figures in the plane. For example, each of the geometric figures in Fig. 6.8 can be paired with a second figure which is congruent to it.

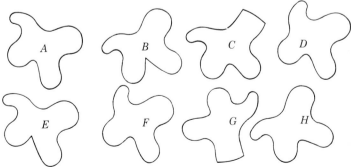

Figure 6.8

In order to gain more familiarity with this definition of congruence we need to ask the question, "Which transformations are isometries?" Certainly translations, rotations, and line reflections are isometries, but are there other types of isometries?

A careful exploration will reveal that there is one additional type, called a **glide reflection**, which is the combination of a translation and a reflection and which is the abstraction of what we called a slide-flip in the preceding chapter.

Theorem 6.1 summarizes the four types of isometries.

Some very nice films exist which describe the isometries. One such film, entitled "Isometries," was developed by mathematicians Seymour Schuster and W. O. J. Moser. It is available from the International Film Bureau, Inc., 332 South Michigan Avenue, Chicago, Illinois 60604.

Theorem 6.1. *A transformation T is an isometry if and only if*

a) T *is a line reflection, or*

b) T *is a translation, or*

c) T *is a rotation, or*

d) T *is a glide reflection.*

There is a close tie between the physical motion of "flipping" explored in Investigations 5.4 and 5.5 and the abstract concept of a line reflection M_p. We draw upon this relationship and the experiences of these investigations to make Theorem 6.1 plausible.

To that end, suppose that A, B, C are three noncollinear points and that A', B', C' are their images under an isometry. We have learned through the experiences in Investigation 5.5 that $\triangle ABC$ can be mapped onto $\triangle A'B'C'$ by a product of at most three line reflections. There are three cases to consider (Fig. 6.9).

These illustrations suggest the following theorem.

Theorem 6.2. *If T is an isometry, then*

a) $T = M_p$, *or*

b) $T = M_q M_p$, *or*

c) $T = M_r M_q M_p$,

for appropriately chosen lines p, q, and r.

By applying Theorem 6.2 we can make Theorem 6.1 plausible by considering products of two or three line reflections.

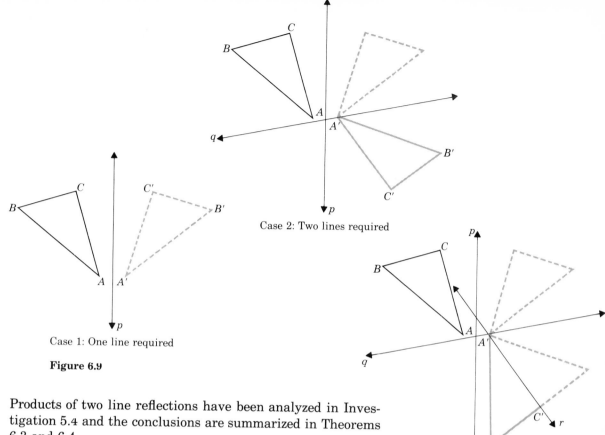

Case 2: Two lines required

Case 1: One line required

Figure 6.9

Case 3: Three lines required

Products of two line reflections have been analyzed in Investigation 5.4 and the conclusions are summarized in Theorems 6.3 and 6.4.

Theorem 6.3. *A transformation T is a translation $S_{AA'}$ if and only if there exists a pair of parallel lines p and q such that $S_{AA'} = M_q M_p$. In that case $p \perp AA'$ and $d(p,q) = \frac{1}{2} AA'$.*

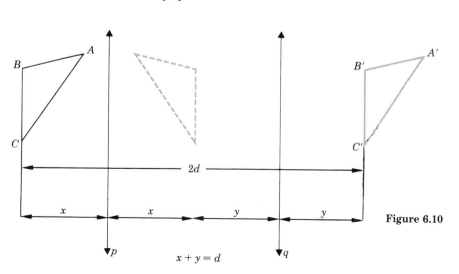

Figure 6.10

$x + y = d$

Theorem 6.4.   *A transformation T is a rotation $R_{0,\alpha}$ if and only if there exists a pair of lines p and q intersecting in O such that $R_{0,\alpha} = M_q M_p$. In that case the measure of the acute angle between lines p and q is $\frac{1}{2}\alpha$.*

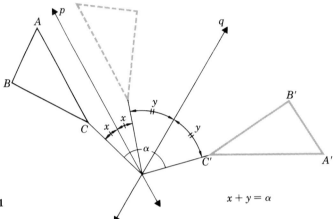

Figure 6.11

$$x + y = \alpha$$

It remains to explore products of three line reflections. In Fig. 6.12 we consider the special case that $p \parallel q$ and $p \perp r$.

The transformation which maps $\triangle ABC$ onto $\triangle A'B'C'$ in Fig. 6.12 is neither a line reflection, a translation, nor a rotation. We call it a glide reflection with axis r. More precisely, a transformation T is a **glide reflection with axis r** if there exist three lines p, q, r, such that $p \parallel q$, $p \perp r$ and $T = M_r M_q M_p$.

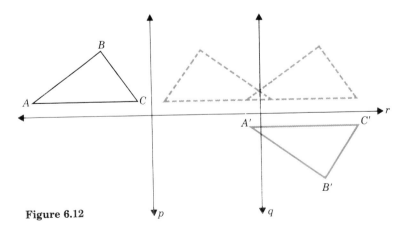

Figure 6.12

Through exercises we see that the product of three reflections may be another line reflection or it may be a glide reflection. Summarizing, the information on the last few pages, taken together, substantiates the claim of Theorem 6.1: Translations, rotations, reflections, and glide reflections are isometries of the plane, and there are no others.

EXERCISE SET 6.4

1. Which pairs of figures are congruent? Why?

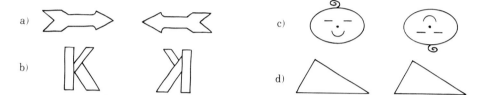

2. For each pair of congruent figures below, describe an isometry which maps one figure onto the other.

3. Given an isometry T, this isometry is called **direct** if a directed circle and its image under T agree in orientation (both clockwise or both counterclockwise). Otherwise T is called an **opposite isometry**. Which of the four types of isometries are direct, and which types are called opposite?

4. Why can we be sure that there is no translation and no rotation which maps $\triangle ABC$ onto $\triangle A'B'C'$ in Fig. 6.12?

5. Suppose lines p, q, and r are concurrent as shown at the right. Find a line s such that $M_r M_q M_p$ is the line reflection M_s.

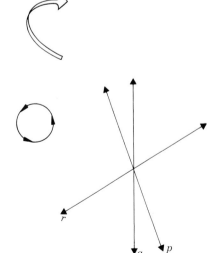

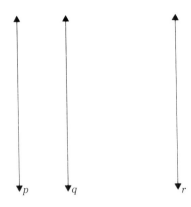

6. Suppose lines p, q, and r are parallel as shown at the left. Find a line s such that $M_r M_q M_p$ is the line reflection M_s.

7. Suppose $\triangle ABC$ and $\triangle A'B'C'$ are congruent, oppositely oriented triangles (i.e. not both clockwise or both counterclockwise).

   a) Convince yourself by construction that the midpoints of AA', BB', and CC' are collinear. Call this line r.

   b) Does M_r map $\triangle ABC$ onto $\triangle A'B'C'$? (It may or may not, depending on your choice for $\triangle ABC$ and $\triangle A'B'C'$.)

   c) If M_r does not map $\triangle ABC$ onto $\triangle A'B'C'$, find a pair of parallel lines p, q with $p \perp r$ such that the glide reflection $M_r M_q M_p$ maps $\triangle ABC$ onto $\triangle A'B'C'$.

8. Each of the transformations below are isometries. Identify which of the four types they are. [*Hint*: Sketch on graph paper a figure and its image.]

   a) $T(x,y) = (x + 3, y + 4)$ b) $T(x,y) = (x+3, -y)$

   c) $T(x,y) = (y,x)$ d) $T(x,y) = (y,-x)$

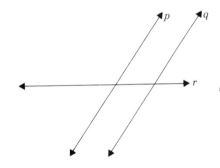

9. Given lines p, q, and r with $p \parallel q$ as shown at the left, construct

   a) $r' = M_q(r)$,   b) $p' = M_q M_p(p)$,

   c) $p' = M_r M_p M_q(p)$.

*10. A rotation about A through $180°$ is called a half turn and is denoted by H_A. Conjecture as a result of construction what type of transformation each of the following are:

   a) $H_A H_B$, if $A \neq B$,

   b) $H_A M_p$, if $A \in p$,

   c) $H_A H_B H_C$, if A, B, and C are three noncollinear points.

*11. Use the distance formula in the definition on page 209 to determine which of the following transformations are isometries.

   a) $T(x,y) = (-x, y+2)$

   b) $T(x,y) = (2x, \frac{1}{2}y)$

   c) $T(x,y) = (x + 3, 2y + 1)$.

*CONGRUENCE OF SPACE FIGURES

Recall that we defined two geometric figures $\mathscr{F}$ and $\mathscr{F}'$ in the plane to be congruent if there exists an isometry of the plane T which maps $\mathscr{F}$ onto $\mathscr{F}'$. Likewise, two geometric figures in space (polyhedra, for example) are congruent if there exists an isometry of space which maps one figure onto the other one. (An isometry of space is a correspondence which associates each point P with a unique point P' such that for each pair of points P, Q, $PQ = P'Q'$.)

We have seen that line reflections are of fundamental importance when considering isometries in the plane since all isometries can be expressed as a product of line reflections (Theorem 6.2). In space the isometry of fundamental importance is the *reflection in a plane*. If α is a plane in space, the **plane reflection** M_α is the transformation which associates each point P with the unique point P' where α is the perpendicular bisector of PP' (Fig. 6.13). The theorem analogous to Theorem 6.2 says that an isometry of space can be written as a product of plane reflections. However, in space up to four plane reflections may be required. We state this theorem.

Theorem 6.5. *If T is an isometry in space, then there exist planes α, β, γ, δ such that*

a) $T = M_\alpha$, or
b) $T = M_\beta M_\alpha$, or
c) $T = M_\gamma M_\beta M_\alpha$, or
d) $T = M_\delta M_\gamma M_\beta M_\alpha$.

The concept of "opposite orientation" is one which most students do not fully appreciate. It is especially important to make cardboard models like those in Fig. 6.13 to convince students that oppositely oriented solid figures cannot be similarly positioned by a flip the way oppositely oriented triangles in the plane can. Martin Gardner's *The Ambidextrous Universe* [39] has an interesting discussion of "orientation."

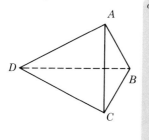
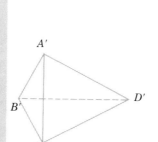

Figure 6.13

We see from this theorem that to determine all types of space isometries, one needs to analyze the products of two, three, and four plane reflections.This analysis would indicate that there are six types of isometries in space.

First consider products of two plane reflections. These two planes could be parallel, or they could intersect in a line l. Analogous to the situation in the plane, if $\alpha\|\beta$, then $M_\beta M_\alpha$ is a translation and if α and β intersect in a line l, $M_\beta M_\alpha$ is a rotation with axis the line l(Fig. 6.14).

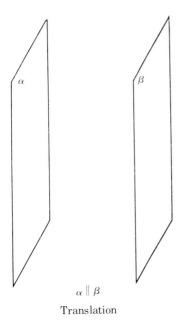

$\alpha \| \beta$

Translation

Figure 6.14

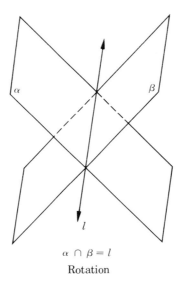

$\alpha \cap \beta = l$

Rotation

When considering the product of three plane reflections, the situation in space is only partly analogous to the situation in the plane. In particular if α, β, γ are three planes with $\alpha\|\beta$ and $\alpha\perp\gamma$, then $M_\gamma M_\beta M_\alpha$ is again a glide reflection. A new situation arises if α, β, γ are planes, with α and β intersecting in l and γ a plane perpendicular to line l. In this case we see that $M_\beta M_\alpha$ is a rotation in line l so that $M_\gamma M_\beta M_\alpha = M_\gamma(M_\beta M_\alpha)$ is a rotation followed by a reflection. This type of transformation is called a **rotatory reflection** (Fig. 6.15)

Finally, the product of four plane reflections for appropriately positioned planes may result in a sixth type of isometry in space. In particular, if α, β, γ, δ are four planes with $\alpha\|\beta$

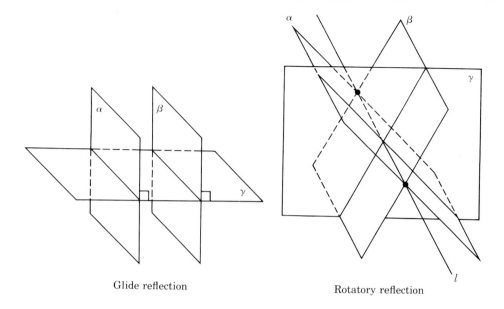

Glide reflection · Rotatory reflection

Figure 6.15

and $\gamma \cap \delta$ is a line l which is perpendicular to the planes α, β then the product $M_\beta M_\alpha$ is a translation and $M_\delta M_\gamma$ is a rotation about line l. Thus the product $(M_\delta M_\gamma)(M_\beta M_\alpha) = M_\delta M_\gamma M_\beta M_\alpha$ is a product of a translation followed by a rotation about a line which is parallel to the direction of the translation. This type of isometry is called a **screw displacement** (Fig. 6.16).

EXERCISE SET 6.5

Which of the six isometries in space is the abstraction of the motion involved in inserting a key in a lock and turning the key?

*SYMMETRY OF POLYHEDRA

In Chapter 4 we examined on the intuitive level the concepts of "plane of symmetry" and "axis of symmetry" for polyhedra. We are now in a position to describe more precisely these two types of symmetry.

Note that the plane α in Fig. 6.17 is a plane of symmetry for the cube. Moreover, the plane reflection M_α maps the cube onto itself.

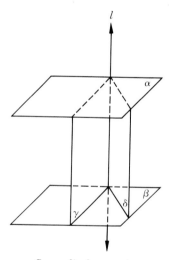

Screw displacement

Figure 6.16

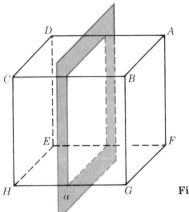

Figure 6.17

That is:

$$M_\alpha(A) = D \qquad M_\alpha(D) = A$$
$$M_\alpha(B) = C \qquad M_\alpha(C) = B$$
$$M_\alpha(F) = E \qquad M_\alpha(E) = F$$
$$M_\alpha(G) = H \qquad M_\alpha(H) = G,$$

and we conclude that $M_\alpha(K) = K$. In other words a plane α is a *plane of symmetry for a polyhedra* $\mathscr{P}$ *if and only if* $M_\alpha(\mathscr{P}) = \mathscr{P}$.

A *line l is an axis of symmetry of a polyhedra* $\mathscr{P}$ if the rotation R_l about axis l maps $\mathscr{P}$ onto itself. That is, $R_l(\mathscr{P}) = \mathscr{P}$.

A cube has axes of rotational symmetry of three different orders (Fig. 6.18). ("Order" is used here as in Chapter 4.)

In the last section we saw that if α and β are planes which intersect in line l, then $M_\beta M_\alpha$ is a rotation with axis $l = \alpha \cap \beta$. This means that if α and β are planes of symmetry for a polyhedra, then $\alpha \cap \beta = l$ is an axis of rotational symmetry for the polyhedra.

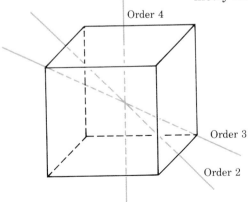

Order 4

Order 3

Order 2   Figure 6.18

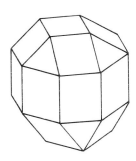

EXERCISE SET 6.6

1. Given a right regular pentagon-based pyramid, plane α, and line l, complete the following:

   a) $M_\alpha(A) =$ $M_\alpha(D) =$ $M_\alpha(C) =$
      $M_\alpha(B) =$ $M_\alpha(E) =$ $M_\alpha(F) =$

   b) $R_{l,\,72}(A) =$ $R_{l,\,72}(B) =$ $R_{l,\,72}(E) =$

2. Given the cube and planes of symmetry α and β as shown in the figure, complete the following:

   $M_\alpha(A) =$ $M_\beta(B) =$ $M_\alpha M_\beta(C) =$
   $M_\beta M_\alpha(F) =$ $M_\beta M_\alpha(C) =$ $M_\alpha M_\beta(F) =$

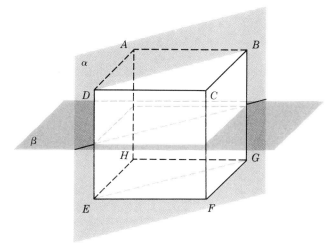

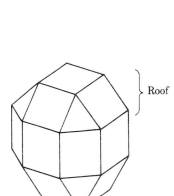

Roof

Rhombicuboctahedron

3. A cube has nine planes of symmetry. Since each pair of planes of symmetry determines an axis of rotational symmetry, and since there are 36 pairs of planes of symmetry, why are there only 13 axes of rotational symmetry for the cube and not 36?

4. A pseudo-rhombicuboctahedron can be derived from a rhombicuboctahedron by cutting off the top "roof" and turning the roof 1/8 turn before glueing it back onto the base.

   Do these two polyhedra possess the same number of planes and axes of symmetry? Explain.

Pseudorhombicuboctahedron

*SYMMETRY PATTERNS IN THE PLANE

In Chapters 3 and 5 we have explored tessellations of the plane and have found many patterns rich in symmetry. In fact, if we consider any of these tessellation patterns as being extended to cover the entire plane, there are four possible types of pattern symmetry. These possibilities are: reflectional symmetry, translational symmetry, rotational symmetry, and glide-reflectional symmetry. For example, a pattern $\mathcal{P}$ possesses *reflectional symmetry* if there is a line reflection M such that $M(\mathcal{P}) = \mathcal{P}$. Figure 6.19 shows a pattern with reflectional symmetry. Each gray line is a line of reflectional symmetry. A pattern $\mathcal{P}$ possesses glide-reflectional symmetry if there exists a glide reflection T such that $T(\mathcal{P}) = \mathcal{P}$. In Fig. 6.20 we see a pattern with glide-reflectional symmetry with the axes of the glide reflections in gray. Similar definitions occur for the other two types of symmetry. As we see in Fig. 6.21, some patterns contain all four types of symmetry.

We shall define two tessellation patterns as belonging to the same symmetry class if they possess identical types of symmetry. For example, the pattern in Fig. 6.22 is in the same class as the pattern in Fig. 6.20 since they both possess axes of glide-reflectional symmetry in two directions and translational symmetry, but possess no rotational or reflectional symmetry. A theorem, difficult to prove and one beyond

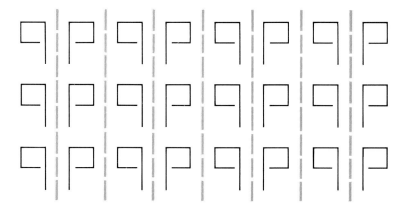

Figure 6.19

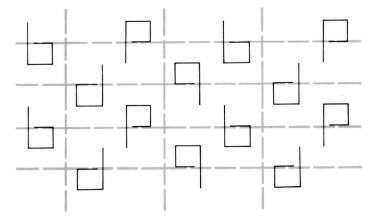

Figure 6.20

the scope of this book, establishes that there are exactly 17
types of symmetry patterns of
the plane. An example of each of
these 17 types is shown in Fig.
6.23.

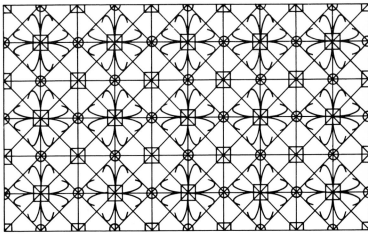

Figure 6.21

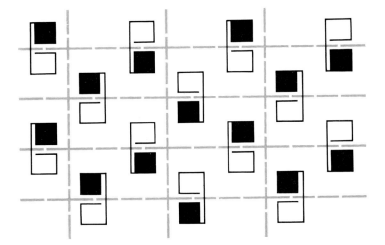

Figure 6.22

(a)

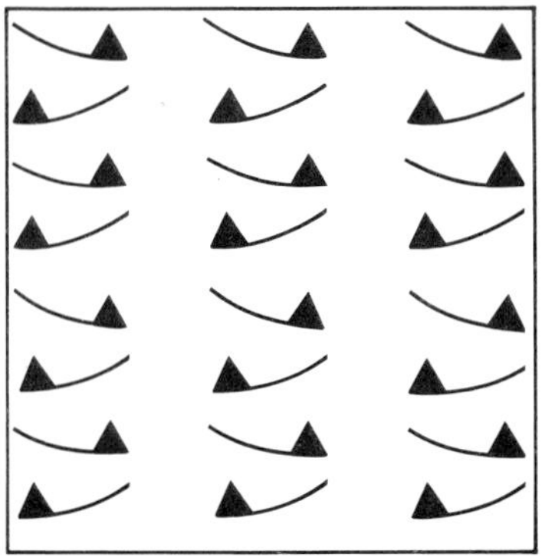

(b)

(c)

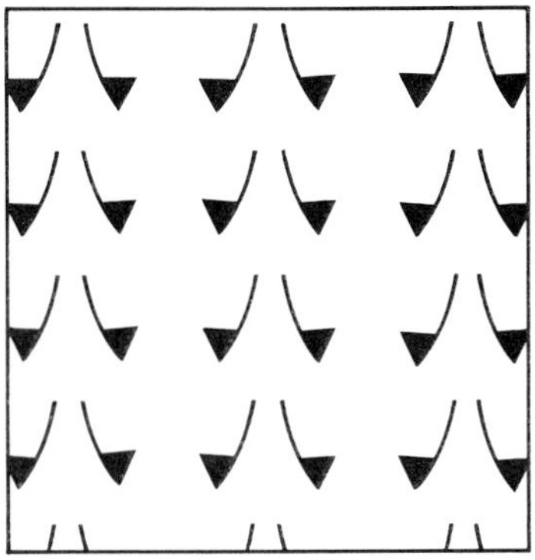

(d)

(e)

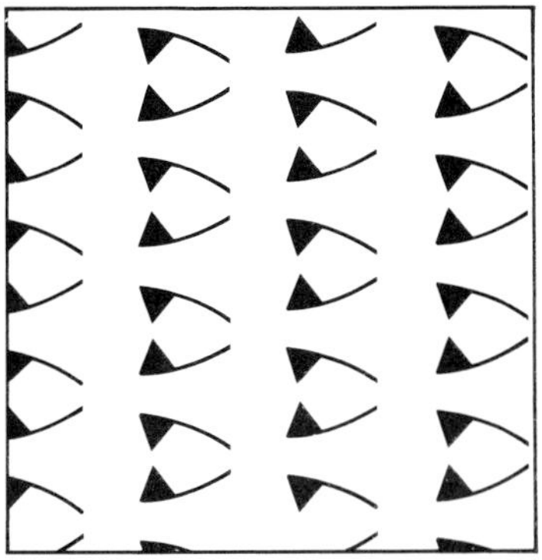

(f)

(g)

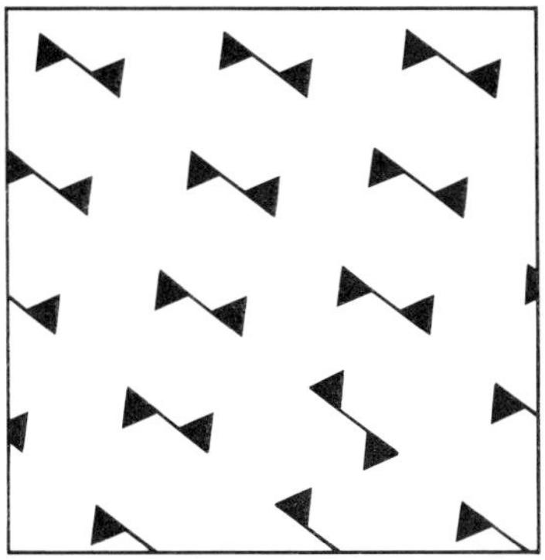

(h)

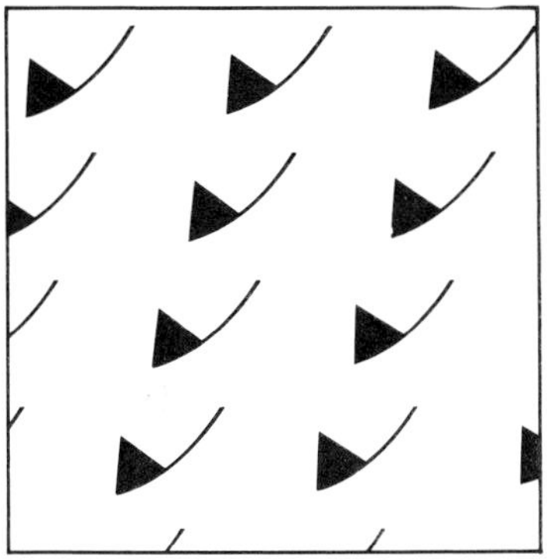

(i)

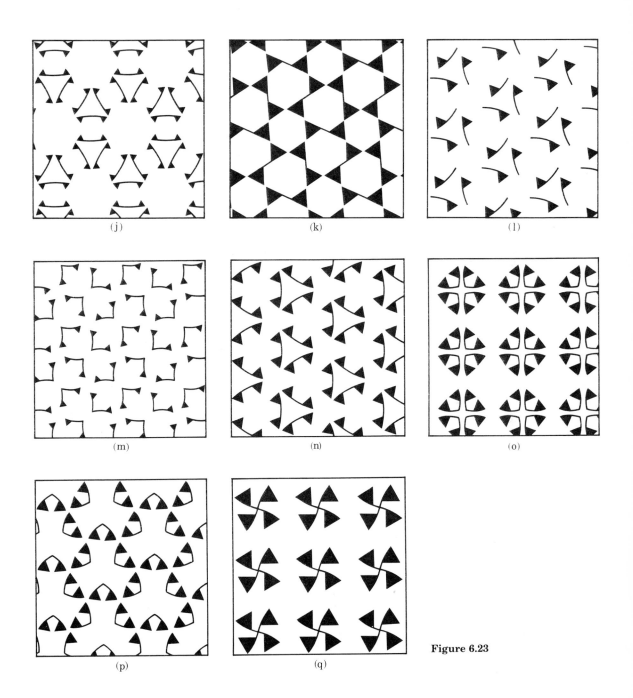

(j)

(k)

(l)

(m)

(n)

(o)

(p)

(q)

Figure 6.23

An exploration of strip patterns can give valuable experience in recognizing reflections, rotations, and translations and discovering combinations of these in an informal way. These ideas may motivate a mathematics project appropriate for junior high or secondary school students. References [20], [34] and [50] provide background information on this topic.

A simple, yet interesting related problem centers around the study of symmetry patterns on a strip. A strip pattern, sometimes called a frieze pattern, is characterized by the fact that there is translational symmetry in exactly one direction. The example in Fig. 6.24 has reflectional symmetry in two different types of lines. One, like line a, the other like line b. On the other hand, the frieze pattern of Fig. 6.25 possesses no reflectional symmetry. It does however, possess rotational symmetry of two types—that like the rotational symmetry about point A and that like the rotational symmetry about point B.

You have an opportunity to construct frieze patterns of your own in the next investigation.

Investigation 6.2

A.

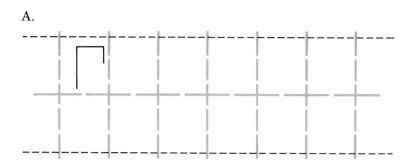

For each part, A and B, complete a frieze pattern beginning with the symbol "Γ" so that each of the gray lines is a line of reflectional symmetry and each point is a center of half-turn symmetry.

B.

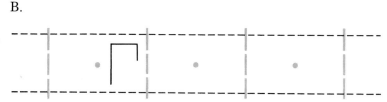

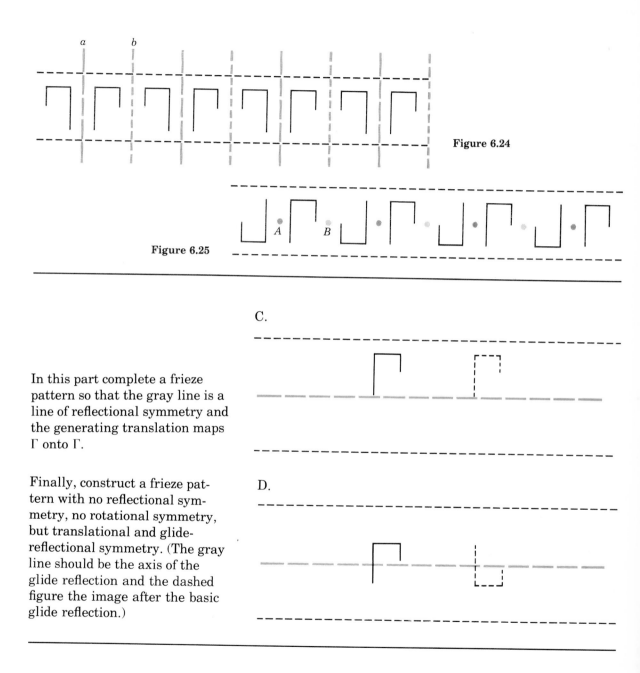

Figure 6.24

Figure 6.25

C.

In this part complete a frieze pattern so that the gray line is a line of reflectional symmetry and the generating translation maps Γ onto Γ.

D.

Finally, construct a frieze pattern with no reflectional symmetry, no rotational symmetry, but translational and glide-reflectional symmetry. (The gray line should be the axis of the glide reflection and the dashed figure the image after the basic glide reflection.)

The patterns in Fig. 6.24 and 6.25, together with the four you constructed in Investigation 6.2, provide examples for six of the seven possible symmetry classes for frieze patterns. All seven types of frieze patterns appear in Fig. 6.26.

EXERCISE SET 6.7

1. Which of the patterns in Fig. 6.23 possess *no* reflectional symmetry?

2. The tessellation in part A of Investigation 5.1 possesses the same type of symmetry as which one of the patterns in Fig. 6.23, (m), (n), or (o)?

3. The frieze pattern which you completed in part A of Investigation 6.2 possesses the symmetry type of which one of the patterns in Fig. 6.26?

4. For each of the frieze patterns below find the one in Fig. 6.26 which possesses the same type of symmetry.

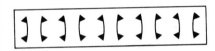

Figure 6.26

BIBLIOGRAPHIC REFERENCES

[4, 158–186], [5, 139–166, 209–218], [17, 30–31, 36–38, 62–70], [19, 504–535], [20, 120–129], [26, 39–47, 96–100], [28, 80–150], [34, 180–188, 207–224], [50], [54], [63], [65], [102, 1–47], [107]

PEDAGOGICAL ACTIVITIES FOR THE TEACHER

1. Study the seventh grade book from the Secondary School Mathematics Curriculum Improvement Study (SSMCIS Unified Mathematics, Course One) Addison-Wesley, 1972, by Fehr, Fey, and Hill. Note particularly the chapters on mathematical mappings and transformations in the plane. List your criteria and evaluate the appropriateness of this material for this grade level.

2. a) A question which has precipitated considerable discussion recently is, "Should the secondary school geometry course be a transformation approach to geometry?" Read articles in the thirty-sixth yearbook of the National Council of Teachers of Mathematics, "Geometry in the Mathematics Curriculum," which speaks to this question, and write a one-page position paper presenting your answer.

    b) In light of your answer to part (a), describe briefly the kind of activities at various levels in the elementary and junior high schools which you think would best prepare a student for the secondary school geometry program which you suggest.

3. Study the approach made by Coxford and Usiskin in their book, *Geometry—A Transformation Approach,* to the topic of congruence.

    a) What postulates do they accept?
    b) What definitions do they make?
    c) What theorems do they prove?
    d) Is the approach based on isometries?
    e) What types of activities in the elementary and junior high schools would provide a good background for a student who will use Coxford and Usiskin's book in tenth grade geometry?

4. Students are often reluctant to begin an individual project in mathematics because

    a) they aren't motivated;
    b) they don't have any references on the topic;

c) they don't see what possible things they could do with the topic.

Prepare a one-page motivational sheet which has the potential to get a student at a chosen level interested in an individual project on the seven different Frieze (Strip) Patterns. Include at least two appropriate references and some suggestions for possible directions which the project might take.

7

Magnification and Similarity

INTRODUCTION

Congruence, a relation studied in the last chapter, is at the very heart of modern technology. Mass production along assembly lines is possible only because we can produce many parts congruent to each other. Figure 7.1 shows how automobile parts, identical in size and shape, are produced and assembled.

Likewise, the relation of "same shape" plays an important role in technology. An integral part of large construction projects, airplanes and buildings for example, is the produc-

Figure 7.1

Teachers of children in early grades can capitalize upon a child's delight in miniature versions of everyday objects. The demand for toy autos and people which are precisely like the real thing show the children's recognition of the correspondence that should exist between part of an object and its model.

Figure 7.2

tion of models and scale drawings in which every aspect of the final object is scaled down by a constant factor. These scaled-down versions of the larger structure (see Fig. 7.2) possess the same shape as the larger object and are essential to providing a full understanding of the intricacies of the actual structure itself.

Photographic enlargement is another method often used to reveal intricate details. It is a useful procedure because the process preserves the shape of objects. For example, in Fig. 7.3 we see a photograph taken at 10,000 feet and an enlargement of a small portion of it.

Figure 7.3

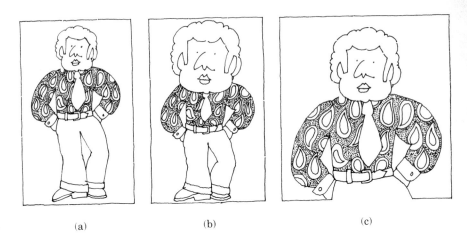

Figure 7.4   (a)   (b)   (c)

We answer two basic questions in this chapter. (1) What precisely is meant by the phrase "same shape"? (2) What characteristics of two figures must be alike before the two figures possess the same shape? For example, *Which one of the three figures in Fig. 7.4 is not the same shape (above the waist) as the other two and why? Which pairs of figures in Fig. 7.5 have the same shape, and why?* Often our intuition leads us to guess that objects are the same shape, but to explain precisely why two figures are not the same shape may be much more difficult.

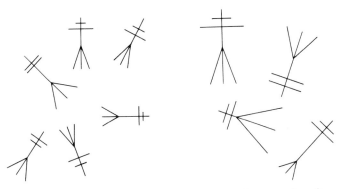

Figure 7.5   Which pairs are the same shape?   Why not the same shape?

MAGNIFICATION

Just as "same size and shape" or "congruence" was described in Chapter 6 in terms of transformations, the concept "same shape," formally called "similarity," can also be described in mathematical terms. An understanding of the concept known as a magnification is essential to a broad understanding of similarity. We begin our study of magnification with an investigation.

Investigation 7.1

Use dot paper or graph paper for this investigation.

A. 1. Multiply the coordinates of points A, B, C by 2 to obtain points A', B', C' respectively, and multiply the coordinates D, E, F, G, H, I, J by ½ to obtain points D', E', ... Show these points on your grid and connect them to form polygons.

   2. Do the points A', B', . . . , J' lie on rays OA, OB, . . . , OJ?

   3. How do the distances OA' and OA compare? How do the distances OB' and OB compare?

   4. Does the $\triangle A'B'C'$ have the same shape as $\triangle ABC$? Do the two heptagons have the same shape?

B. Conjecture what the results in step 2 would be if all coordinates were multiplied by 3; by 4; by ½. Check your conjecture through drawings on dot paper.

C. Draw on dot paper a picture of your own choosing. Draw the figure resulting from multiplying all coordinates (relative to some coordinate system) by 2.

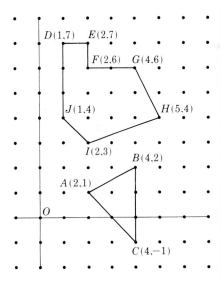

We have seen in this investigation that when all coordinates are multiplied by a fixed constant k, the resulting figure has the same shape as the original figure. The perimeter of

Students enjoy making magnifications of pictures, using coordinates on graph paper. A picture such as the cartoon character below is placed on a small grid graph paper and key points are labeled. These points are plotted on larger grid paper and the figure is reproduced. This magnification can also be done by doubling or tripling the numbers in the ordered pair for each point.

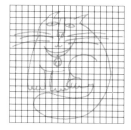

the resulting figure is k times the perimeter of the original figure, and the area of the resulting figure is k^2 times the area of the original figure. These observations are accurate and are the result of the following basic fact: If the coordinates of the points A, B, C are multiplied by the positive number k,

$$OA' = k \cdot OA, \quad OB' = k \cdot OB,$$

and

$$OC' = k \cdot OC.$$

This observation motivates the concept we call a magnification with center O and scale factor k.

A **magnification with center** O **and scale factor** $k > 0$, denoted $M(O,k)$ maps each point $P \neq O$ to the unique point P' on ray $\overrightarrow{OP}$ where $OP' = k \cdot OP$, and maps point O to itself. The point P' is called the **image of** P **under the magnification** $M(O,k)$. (Note that if $k < 1$, $M(O,k)$ actually has a shrinking effect rather than a magnifying one.)

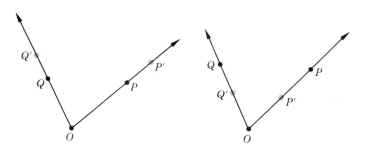

$$\frac{OP'}{OP} = \frac{OQ'}{OQ} = k, k > 1 \qquad \qquad \frac{OP'}{OP} = \frac{OQ'}{OQ} = k, k < 1 \qquad \textbf{Figure 7.6}$$

If a pen light is used as the point and a figure is held perpendicular to the penlight ray, the magnification of the figure in three-space appears in the shadow of the figure on the wall. Children can experiment with the effect of the proximity of the point to the figure on the size of the magnification.

Note that the definition of magnification does not refer to a coordinate system even though the definition was motivated by using one. The point O can be any point in the plane, and is not restricted to the origin of a coordinate system.

By finding the image of each point P in a set, we obtain what we call the image set. In Fig. 7.7 we see two figures (in black) and their images (in gray) under the magnification $M(O, 3/2)$. In the following exercises we continue to explore magnifications.

$\bullet$
O

Figure 7.7

EXERCISE SET 7.1

Use dot paper or graph paper for these exercises.

1. a) Draw the image of $\triangle ABC$ under the transformation $M(O_1, 2)$.

   b) Draw the image of $\triangle ABC$ under the transformation $M(O_2, 1/2)$.

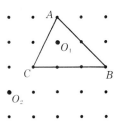

2. a) Draw the circle C_2 whose image under the transformation $M(O_1, 2)$ is the circle C_1.

   b) Draw the circle C_3 whose image under $M(O_2, 2)$ is the circle C_1.

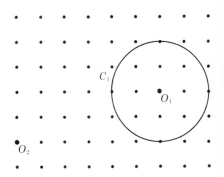

3. Given the two figures on the right:

   a) Find a center O and a scale factor k such that the transformation $M(O, k)$ maps the small figure to the large figure.

   b) If $M(O, k')$ maps the large figure onto the small figure, how are k and k' related?

4. Convince yourself through drawings on dot paper that if $M(O, k)$ is any magnification, then

   a) the image of line l under $M(O, k)$ is a line parallel to l,

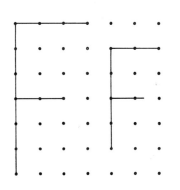

b) the image of a ray is another ray,

c) the image of an angle is an angle.

5. Let $\angle ABC$ be any angle. If $\angle A'B'C'$ is the image of $\angle ABC$ under a magnification $M(O,k)$, how does the measure of these two angles compare?

6. Does there exist a magnification which maps $\triangle ABC$ onto $\triangle A'B'C'$? If so, find its center. If not, why not?

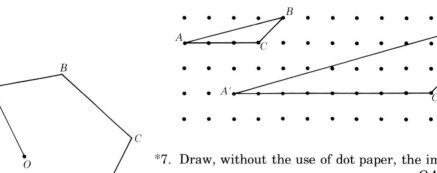

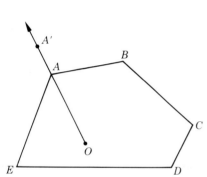

*7. Draw, without the use of dot paper, the image of the pentagon $ABCDE$ under $M(O,k)$ if $k = \dfrac{OA'}{OA}$. Describe the procedure that you followed.

*8. The following game can be played with a partner or as solitaire. Select any pair of points A and B on a sheet of dot paper.

Player 1 begins at point A and draws a horizontal or vertical line segment to some other dot. *Player 2* is obligated to respond by drawing a segment in a parallel direction through twice the distance. *Player 1* continues by drawing another horizontal or vertical segment and *Player 2* responds accordingly. The game continues in this fashion.

Object: Select a positive integer $k > 2$. *Player 1* tries to find a sequence of moves so that *Player 1* and *Player 2* end at the same dot after each player has taken exactly k moves. (You should assume that your sheet of dot paper is of unlimited size.)

Sample Game

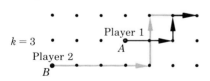

Question: Can the object of the game be met no matter what value of k you select and no matter what pair of points A and B you begin with? Explain.

*9. There are many variations of the game in Exercise 8. Select some variation so as to create a slightly different game.

PROPERTIES OF MAGNIFICATIONS

Through the experiences of Investigation 7.1 and the exercises in the previous section, the reader may have already discovered many properties of magnifications. In this section we summarize some of these properties, concentrating on geometric properties which are preserved under magnifications.

1. *Betweenness is preserved.* (1) If B is between A and C, then B' is between A' and C'.

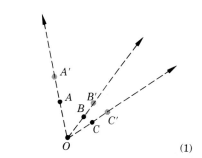

(1)

2. *Lines are preserved.* (2) If l is a line, then the set l' of all images of points of l is also a line. In fact l' is always parallel to l.

3. *Rays and Segments are preserved.* (3) The image of a ray $\overrightarrow{AB}$ is a ray $\overrightarrow{A'B'}$, and the image of a segment $\overline{CD}$ is a segment $\overline{C'D'}$.

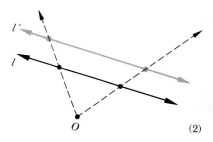

(2)

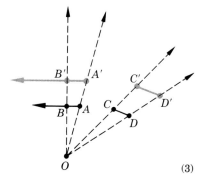

(3)

4. *Angles and their measure are preserved.* (4) Since the image of a ray is a ray, it follows that the image of an angle is an angle. The measure of an angle and its image are equal. For example $m(\angle ABC)$ is equal to $m(\angle A'B'C')$.

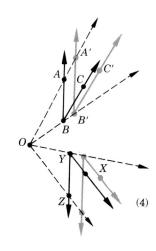

(4)

5. *Polygons are preserved.* Since the image of a segment is a segment, it follows that the image of a polygon is a polygon.

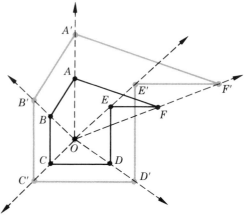

6. *Ratio of distances is preserved.* The ratio AB/BC is equal to the ratio $A'B'/B'C'$. In general, the ratio of the lengths of two segments is equal to the ratio of the lengths of the image segments. The above equality of ratios can be written:

$$\frac{AB}{BC} = \frac{A'B'}{B'C'}.$$

The sixth property says that the ratio of two segments of a figure is equal to the ratio of the corresponding pair of segments in the image figure. That is, given points A, B, C, and D,

$$\frac{AB}{CD} = \frac{A'B'}{C'D'}.$$

From this equation we can derive the equality

$$\frac{A'B'}{AB} = \frac{C'D'}{CD}$$

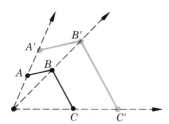

which says that the ratio of a segment and its image is the same beginning with AB as when beginning with CD. It can be shown that this latter ratio is equal to the scale factor k of the magnification. In other words, given a figure and its image under a magnification $M(O, k)$, the distance $A'B'$ is k times the distance AB for each pair of points A and B.

In his interviews, Piaget has observed that a fullblown concept of ratio does not usually appear until well into the formal operations stage, around ages 11 and 12. Teachers of younger children should create experiences which prepare for this concept acquisition. Teachers of older students should utilize the ideas of ratio to study similar triangles and to solve other problems in geometry.

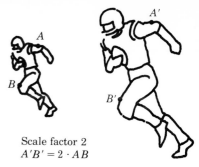

This fact illustrated in Fig. 7.8 ensures that *any figure and its image under a magnification are the same shape.*

Scale factor 2
$A'B' = 2 \cdot AB$

Figure 7.8

EXERCISE SET 7.2

1. Find the center of the magnification which maps A to A' and B to B'. (1)

(1)

2. Given that the magnification which maps B to B' has its center on line l, describe how to construct the image point A' under the magnification. (2) *Hint:* $\overleftrightarrow{AB} \parallel \overleftrightarrow{A'B'}$.

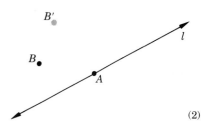

(2)

*3. Given a magnification which maps A to A' and B to B', construct the image of point C under this magnification. (3)

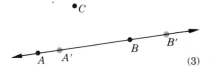

(3)

4. Given a parallelogram $AA'B'B$, explain why there can be no magnification which maps A to A' and B to B'. (4)

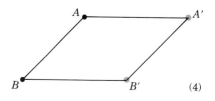

(4)

5. Suppose $ABCD$ is a parallelogram. If A', B', C', D' are the images of A, B, C, D under a magnification, explain why we know that quadrilateral $A'B'C'D'$ is a parallelogram. (5)

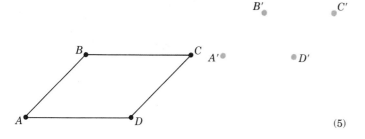

(5)

6. Given $\triangle ABC$ with altitude $\overline{AM}$, if A', B', C', and M' are images of A, B, C, M under a magnification, explain why we can assert that $A'M'$ is an altitude of $\triangle A'B'C'$.

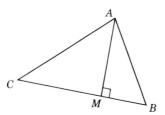

7. Suppose a quadrilateral $ABCD$ is a kite and that A', B', C', D' are the images of A, B, C, D under a magnification. Explain why quadrilateral $A'B'C'D'$ is also a kite.

SIMILARITY

We have seen in the last section that any geometric figure and its image under a magnification possess the same shape. The converse situation, however, is not true. Two figures may have the same shape without one being the image of the other under a magnification. For example, by twisting either the original or the image figure in Fig. 7.8 (see Fig. 7.9) we do not destroy the quality of possessing the same shape. However, since a line and its image under a magnification are parallel, one figure cannot be mapped onto the other by a magnification. So it is apparent that to study the relation "same shape," we must consider combinations of motions and magnifications. We shall do this in investigation 7.2.

Figure 7.9

Investigation 7.2

For each pair of similarly shaped figures:

1. Find a center of a turn or a line of reflection (whichever is required) and a center O for a magnification such that the motion followed by the magnification maps Figure 1 onto Figure 2. In how many ways can this be done?

2. Repeat this process showing that Figure 1 can be mapped onto Figure 2 by a magnification followed by a motion. Convince yourself by completing accurate constructions that this can be done in more than one way.

(Protractor, straight edge, and tracing paper should be an aid for this investigation.)

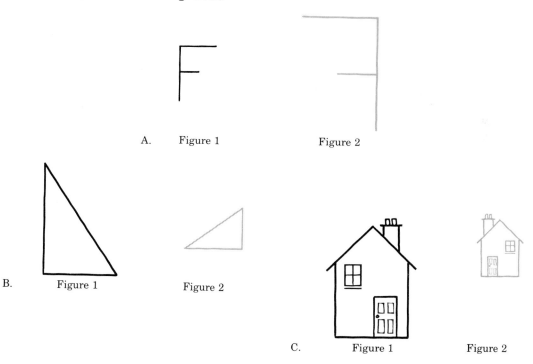

A.   Figure 1   Figure 2

B.   Figure 1   Figure 2

C.   Figure 1   Figure 2

The experience of Investigation 7.2 indicates that given two figures, one the enlargement of the other, after one is repositioned appropriately by a turn or flip, the first can be magnified to coincide with the second. This observation is the basis for the following definition of similarity.

Definition. *Two geometric figures $\mathcal{F}$ and $\mathcal{F}'$are similar if there exists a motion followed by a magnification which maps one figure onto the other figure.*

Note that this definition of similarity is not restricted to only triangles, nor even to polygons. It applies to all types of geometric figures. We see from this definition that exploring products of motions and magnifications is in essence exploring the relation of similarity.

EXERCISE SET 7.3

1. For each part below, draw on dot paper a polygon similar to the given one by performing on the given polygon:

   a) $M(O, 3/2)$,

   b) a counterclockwise turn through 90° followed by $M(O, 2)$,

   c) a flip in line p followed by $M(O, 2)$, and

   d) a flip in line q followed by $M(O, 2)$.

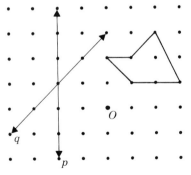

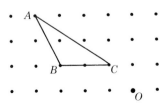

2. a) Draw on dot paper $\triangle A'B'C'$, the image of $\triangle ABC$ under a 90° clockwise turn about O followed by the magnification $M(O, 2)$.

b) Draw on the same dot paper $\triangle A''B''C''$ the image of $\triangle ABC$ under the magnification $M(O, 2)$ followed by a 90° clockwise turn about O.

c) Are $\triangle A'B'C'$ and $\triangle A''B''C''$ the same triangle? Are they similar?

3. Try other examples like Exercise 2 and search for a generalization. Is the image of a figure under a turn about point O followed by a magnification $M(O, k)$ the same set as the image of the figure under the magnification $M(O, k)$ followed by a turn about point O? Always? Only sometimes? Or never?

4. a) Draw on dot paper the image $A'B'C'D'$ of quadrilateral $ABCD$ under the magnification $M(O_1, 2)$, and in turn draw the image $A''B''C''D''$ of this enlarged quadrilateral under $M(O_2, \frac{1}{2})$.

   b) What single motion would map quadrilateral $ABCD$ onto quadrilateral $A''B''C''D''$?

   c) Generalize! Conjecture a theorem describing the motion $M(O_1, k)$ followed by $M(O_2, 1/k)$.

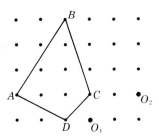

5. Look around your environment. Do you see examples of congruence? Do you see examples of similarity? Which relation seems to be most prevalent in the physical world, congruence or similarity?

*6. Suppose a cube, 1 meter on an edge, is filled with 1 cm spheres. Suppose a second cube, 2 meters on a side, is filled with 2 cm spheres. In both cases consider the ratio of the volume of the spheres to the volume of the entire box. In which case is this ratio the greatest? (Assume that both boxes contain the same number of spheres.)

SIMILARITY OF POLYGONS

Studying "similarity" or "same shape" in terms of motions and magnifications as we did in the last section is intuitively appealing and provides a broad understanding of this concept. The process of photographic enlargement is a helpful and accurate physical model for thinking about similarity. On the other hand, there are some very practical problems for which it is more convenient to describe similarity in terms of angle measures and ratios of segments. For example, suppose a draftsman is asked to complete an enlarged drawing with scale factor of 9/4 of the floor plan in Fig. 7.10. What types of measurements must he make and how many? Or consider the job of surveying a field and producing a map like the one appearing in Fig. 7.11. What step by step procedure does the surveyor use in constructing this map? The purpose of this section is to provide a description of similarity for polygons in terms of angle measures and ratios of segments.

A pantograph is a useful mechanical device for producing a figure similar to a given figure. Upper-level students might enjoy constructing a pantograph and proving that the figure produced is, indeed, similar to the original figure. Information on a pantograph can be found on p. 127 in the NCTM Eighteenth Yearbook, entitled *Multi-sensory Aids in the Teaching of Mathematics,* or in *Geometrical Tools, A Mathematical Sketch and Model Book* (Educational Publishers, 1949) by Robert C. Yates.

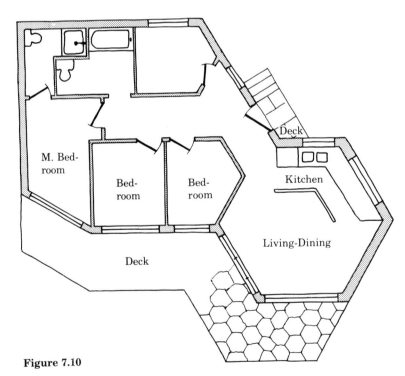

Figure 7.10

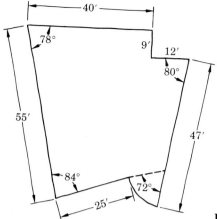

Figure 7.11

In the last section we established that two polygons are similar if there exists a motion followed by a magnification which maps one of the polygons onto the other. This process establishes a one-to-one correspondence $\{A, B, C, \ldots\}$ $\leftrightarrow \{A', B', C', \ldots\}$ between the vertices of the two polygons. Since a magnification preserves angle measure and the ratio of distances, we see that for similar polygons:

a) corresponding vertex angles have equal measure, and

b) the ratio of lengths of corresponding sides is equal to the scale factor of the associated similarity transformation.

These facts are illustrated for a pair of pentagons and a pair of triangles in Fig. 7.12 and 7.13.

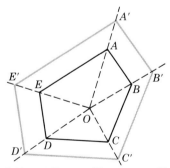

$m\angle A = m\angle A', \ m\angle B = m\angle B', \ \ldots \ m\angle E = m\angle E'$

$$\frac{A'B'}{AB} = \frac{B'C'}{BC} = \frac{C'D'}{CD} = \frac{D'E'}{DE} = \frac{A'E'}{AE} = \frac{OA'}{OA}$$

Figure 7.12

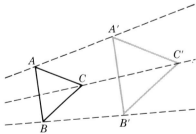

$m(\angle A) = m(\angle A'); \ m(\angle B) = m(\angle B'); \ m(\angle C) = m(\angle C')$

$$\frac{A'B'}{AB} = \frac{B'C'}{BC} = \frac{A'C'}{AC} = \frac{OA'}{OA}$$

Figure 7.13

The converse situation is also true. If there exists a one-to-one correspondence between vertices of two polygons such that

a) corresponding vertex angles have equal measures, and

b) the ratios of corresponding sides are equal to each other,

then the two polygons are similar. We shall illustrate this assertion for the pair of pentagons in Fig. 7.14. The one-to-one correspondence $\{A,B,C,D,E\} \leftrightarrow \{A',B',C',D',E'\}$ between these two pentagons satisfies the conditions:

a) $m(\angle A) = m(\angle A'), \ldots, m(\angle E) = m(\angle E')$

and

b) $\dfrac{A'B'}{AB} = \dfrac{B'C'}{BC} = \cdots = \dfrac{A'E'}{AE}.$

We shall show that these pentagons are similar by describing a motion followed by a magnification which maps (A,B,C,D,E) onto (A',B',C',D',E').

Since $m\angle A = m\angle A'$, there exists a motion (turn or flip) which maps (A, B, C, D, E) to the position (A', B'', C'', D'', E'') shown in Fig. 7.15.

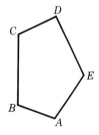

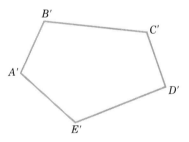

Figure 7.14 $\{A,B,C,D,E\} \leftrightarrow \{A',B',C',D',E'\}$

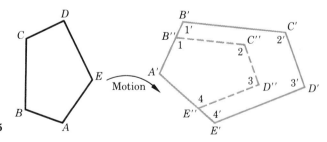

Figure 7.15

Since $\angle 1 = \angle 1', \angle 2 = \angle 2', \quad \angle 3 = \angle 3', \quad \angle 4 = \angle 4'$, it follows from elementary properties of parallel lines that $\overline{B'C'} \| \overline{B''C''}$, $\overline{C'D'} \| \overline{C''D''}$, and $D'E' \| \overline{D''E''}$. With these facts, together with the fact that the ratios of corresponding sides are equal, it can be shown that the magnification $M(A', k)$ with $k = \dfrac{A'B'}{A'B''}$ maps $(A', \quad B'', \quad C'', \quad D'', \quad E'')$ onto (A', B', C', D', E'). Consequently the motion, followed by the magnification on $M\!\left(A', \dfrac{A'B'}{A'B''}\right)$, maps pentagon (A, B, C, D, E) onto pentagon (A', B', C', D', E') and these two pentagons must be similar. This argument can be repeated for any other pair of similar polygons. We summarize this discussion by stating a theorem.

Theorem. *Two polygons are similar if and only if there exists a one-to-one correspondence between the vertices of the polygons such that:*

a) *corresponding angles possess equal measures, and*

b) *ratios of corresponding segments are equal.*

This theorem indicates, if we assume accurate measurements of angles and sides are possible, that given any polygon and any ratio factor, a polygon similar to the given one can be drawn.

EXERCISE SET 7.4

1. For each pair of similar polygons below, write down a one-to-one correspondence between the vertices which satisfies the conditions of the above theorem.

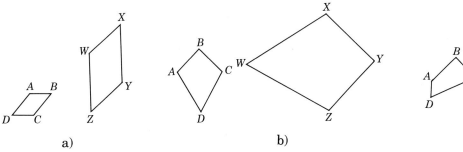

a)   b)   c)

2. Each of the following pairs of triangles are similar. Find all missing measures of the sides.

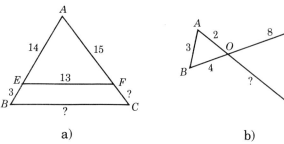

a)   b)

3. The line of sight across a 2-ft and a 3-ft stick crosses the top of a tree. The 3-ft stick is 100 ft from the base of the tree, and the 2-ft stick is 102 ft from the tree. How tall is the tree?

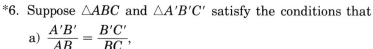

4. A camper wishes to estimate the width of a river. He stands at point A directly across from a large tree and drives a stake at point B, on the bank of the river. Next the camper places a stake at point C along his line of sight to a large rock on the opposite side of the river so that $\overline{BC}$ is perpendicular to $\overline{AB}$. The camper measures $\overline{AB}$ to be 3 m, $\overline{BC}$ to be 1.7 m, and the distance from the tree to the rock to be 75 m. How wide is the river?

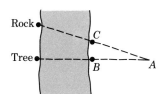

5. Suppose $\triangle ABC$ and $\triangle ADE$ are related as in the figure at the left. Use the definition of similarity of geometric figures to show that $\triangle ABC$ and $\triangle ADE$ are similar.

[*Hint:* Consider the magnification $M(A, \dfrac{AB}{AD})$.]

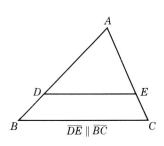

*6. Suppose $\triangle ABC$ and $\triangle A'B'C'$ satisfy the conditions that

a) $\dfrac{A'B'}{AB} = \dfrac{B'C'}{BC}$,

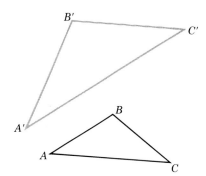

b) $m(\angle B) = m(\angle B')$.

Show that these triangles are similar by showing that there exists a motion followed by a magnification which maps $\triangle ABC$ onto $\triangle A'B'C'$.

(Conditions (a) and (b) in this problem are known as the SAS condition for similarity of triangles since they involve two sides and the included angle of a triangle.)

7.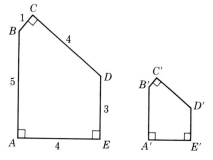

Given are these two pentagons with angles A, C, E, A', C', E' all right angles.

a) If the only additional information is that $m(\angle B) = m(\angle B')$, are these pentagons necessarily similar?

b) If in addition to the information above, you know that $A'B' = 2\frac{1}{2}$ and $A'E' = 2$, are these two pentagons necessarily similar?

*8. Suppose $\triangle ABC$ and $\triangle A'B'C'$ are related by $m\angle A = m\angle A'$, $m\angle B = m\angle B'$, $m\angle C = m\angle C'$.

Show that these triangles are similar by showing that there exists a motion followed by a magnification which maps $\triangle ABC$ onto $\triangle A'B'C'$.

(The condition in this exercise is known as the AAA condition for similarity since it involves the three angles of the triangle.)

9. Begin with segment $A'B'$ and use a ruler and protractor to draw a figure $A'B'C'D'E'F'G'$ which is similar to figure $ABCDEFG$ shown here.

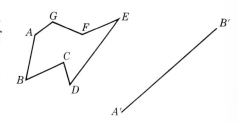

*10. Find (without measuring) the lengths of the segments $\overline{AX}$, $\overline{XB}$, $\overline{CX}$, $\overline{XD}$.

*11. a) Name as many different convex shapes as you can in the figure below. Are any of them hexagons?

   b) How many similar figures can you find?

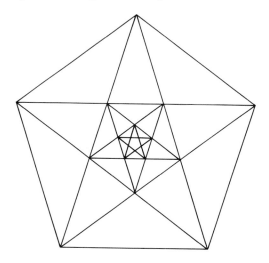

BIBLIOGRAPHIC REFERENCES
[5, 189–201], [6, 149–156], [26, 67–74, 96–100], [28], [34], [54]

PEDAGOGICAL ACTIVITIES FOR THE TEACHER

1. Create a concept card using examples and nonexamples to help a student understand the concept of similar polygons. Try the concept card with a student and revise it to improve its effectiveness.

2. A useful outline for preparing a lesson plan is as follows:

   I. Preparation (reviewing basic ideas and providing motivation for the next phase).

   II. Investigation (an exploratory phase in which the student is actively involved with physical materials, often with a group of other students and one in which the student takes major responsibility for exploring a central idea of the lesson).

III. Discussion (a phase in which the result of the investigation is discussed and one in which the teacher shapes the concept of the idea to be developed).

IV. Utilization (a phase in which the student uses the ideas developed in the two previous phases).

V. Extension (a phase in which both remedial and enrichment needs are met).

Use this outline and prepare a lesson plan for teaching scale drawing at an appropriate level. Include in your lesson plan:

a) an idea which would arouse the students' curiosity about the lesson during the preparation phase;

b) an activity card which could be used by a group of students during the investigation phase, to explore basic ideas;

c) a list of at least five questions you would ask the students during the discussion phase.

d) at least two sample problems from a ditto sheet the students could use during the utilization phase.

e) an activity which could be used for enrichment during the extension phase.

3. Seek a reference on the pantograph and write a set of instructions for making and using this device at a grade level of your choice.

4. Make a sequence of overhead projectuals which could be used to develop some idea of magnification or similarity. You may want to develop these ideas with young children or with junior high or secondary students. Attempt to include an element of the discovery approach in your projectual presentation.

5. Make a list of household items which could be used in a lesson to illustrate the concepts of similarity and non-similarity. Plan a brief student activity or teacher demonstration using the materials on your list.

8

Topology

Investigation 8.1

Take a strip of paper 10″ by 1″, give one end a half-twist, and join the ends together with tape.

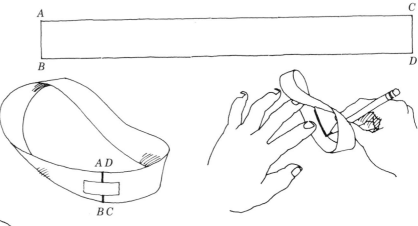

1. Color the edge of the Moebius band until you return to the beginning point.

   What does this tell you about the number of edges on a Moebius band?

2. Punch a hole in the surface of the band with a sharp object. Label the entrance of the point X and the exit of the point Y.

   Draw a continuous line (without crossing the edge) from X to Y. What does this tell you about the number of sides to the band?

3. Along the middle of the band draw a line; predict the outcome, then cut the band in half by cutting along this line.

INTRODUCTION

Paul Bunyan and his mechanic, Ford Fordsen, had started to work a uranium mine in Colorado. The ore was brought out on an endless belt which the manufacturers had made all in one piece, without any splice or lacing, and they had put a half-twist in the return part so that the wear would be the same on both sides. After several months of operation, Paul decided he needed a belt twice as long and half as wide. He told Ford Fordsen to take his chain saw and cut the belt in two lengthwise.

"That will give us two belts," said Ford Fordsen. "We'll have to cut them in two crosswise and splice them together."

"No," said Paul. "This belt has a half twist — which makes it what is known in geometry as a Moebius strip."*

In Investigation 8.1 we shall explore the Moebius band and discover why Paul said that no splicing would be necessary.

As you discovered in Investigation 8.1, a Moebius band is a surface with exactly one edge and one side. Hence, when it is cut in half lengthwise, it remains in one piece. (Notice that in step 2 of the investigation, the tip of the object which punctured the surface exited the same side which it entered.)

The Moebius band is only one of the interesting and curious surfaces which can easily be made with paper, scissors, and tape. In the following exercises we shall further explore the Moebius band as well as other surfaces.

Topology provides a rich source of interesting activities for students at all levels. Two books containing interesting activities are *Topology, The Rubber Sheet Geometry* (Webster, 1960) by Donovan Johnson and William Glenn, and *Visual Topology* (American Elsevier Publishing Company, 1969) by W. Liepzmann.

EXERCISE SET 8.1

1. In a Paul Bunyan story which appears in *Topology* by D. A. Johnson and W. H. Glenn, Loud Mouth Johnson, who was described as Public Blow-Hard Number One, lost $1000 when he bet Ford Fordsen that Paul's belt would end up in two pieces. Loud Mouth subsequently stumbled onto Paul in the process of cutting the belt a sec-

---

* Johnson, D. A., and W. H. Glenn, *Topology*. (St. Louis: Webster, 1960, p. 12.)

ond time to obtain a belt four times the length and one-fourth the width. When Loud Mouth was informed that Ford had gone into town to get splicing material, he informed Paul, having learned through the loss of the $1000, that no splicing material would be necessary.

Make another paper Moebius band and discover whether Loud Mouth was correct.

2. Make another Moebius band and draw a line one third of the width from an edge. Cut along this line, but predict the result before cutting. Were you correct?

3. A Moebius band is made by giving a strip of paper one half-twist. Complete the following table and see what happens when bands constructed with a different number of twists are cut down the middle.

Number of half-twists	Number of sides before cut	Number of edges before cut	Number of sides after cut	Number of edges after cut
0	2		2	1
1				1
2		1	2	2
3		1	2	2

4. a) Which of the following surfaces are *one* sided?

   b) How many edges does each surface have?

  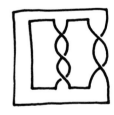

5. Figure out a way of constructing with paper, scissors, and tape each of the surfaces in Exercise 4.

*6. In *The Unexpected Hanging,* Martin Gardner includes a chapter on Borromean rings. The three rings pictured below are Borromean rings.

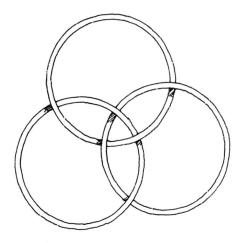

Two of the three surfaces below have Borromean ringed edges. Which two?

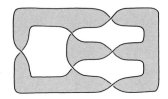 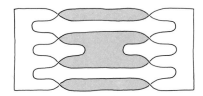 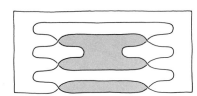

*7. Construct bands with 0,1,2,3, and 4 half-twists (the Moebius band is constructed by one half-twist). Cut each of these bands in half lengthwise to see how many separate bands (which may be interlocking) you obtain. Can you guess at a general pattern.

*8. Construct bands with 0, 1, 2, 3, and 4 half-twists. Cut each of these bands in half lengthwise and check the number of twists in the resulting band or bands. (You may need to cut across the resulting band and count the number of half-twists required to untwist the band.)

*9. Construct bands with 1,3, and 5 half-twists and cut each in half lengthwise. The resulting band will be knotted. Compare the nature of the resulting knots. Are some more knotted than others? Can you find a general pattern?

TOPOLOGICAL EQUIVALENCE

In an article by Piaget on "How Children Form Mathematical Concepts," (*Scientific American,* November 1953), it is asserted that the first geometric notions that a child learns appear to be those of a topological character. For example, at the age of three, a child cannot tell the difference between squares and triangles, but can distinguish between open and closed figures. Topological ideas developed early by children include proximity, separation, order, enclosure, and continuity. These ideas are more carefully described in the book, *How Children Learn Mathematics* (Macmillan, 1970) by Richard W. Copeland, in the chapter entitled, "How a Child Begins to Think about Space."

Topology is a relatively new field in geometry which has come into full bloom since 1920. A topologist, who is a mathematician whose area of special interest is topology, has often been described in the popular press as one who cannot tell the difference between a coffee cup and a donut. How could anyone with the reputed intelligence of mathematicians have such a problem? We hope to shed some light on this apparant paradox as we proceed through this chapter.

When studying congruence and similarity we were concerned with distances between points, lengths of segments, and angle measures—all numerical quantities. In this chapter we shall study geometric concepts which can be described without reference to numbers. Furthermore, these concepts deal with properties of curves and surfaces which are maintained as they are stretched and distorted. The property "has no ends" for a circle is an example of such a concept. Figure 8.1 illustrates that this property of a circle is not lost as the circle is distorted into a triangle.

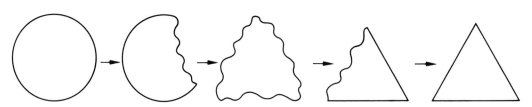

Figure 8.1

It is through this type of distortion that many physical settings can profitably be reduced to a diagram on paper. For example, in Fig. 8.2 we see a partial schematic for a radio receiver. In Fig. 8.3 we see a diagram of the London Underground system. Each figure represents a distortion of a physical situation which preserves the basic structure of the situation.

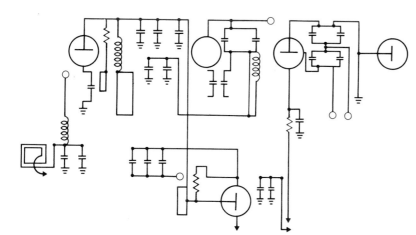

Fig. 8.2 Partial schematic for a radio receiver.

Fig. 8.3 Diagram of part of the London Underground system.

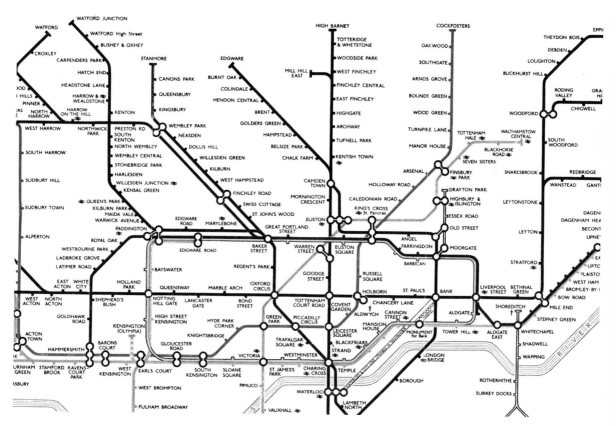

Our objectives for the remainder of this chapter are two-fold. First, we shall describe more carefully these types of distortions, which we shall call topological transformations; and second, we shall discover some of the geometrical properties of surfaces and line figures which remain unchanged by these transformations. The first of these objectives is considered in the remainder of this section.

Figure 8.4

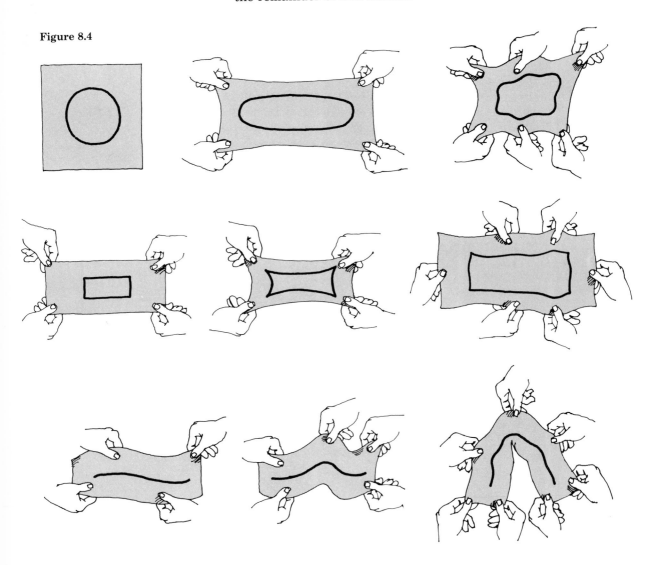

A topological transformation will distort, stretch, shrink, or bend a curve or surface. It will not weld, cement, or join together, and with one type of exception, it does not tear or cut. A stretching rubber sheet is a helpful physical model of a topological transformation. Figure 8.4 illustrates how a circle, a rectangle, and a line segment can be distorted by topological transformations. In fact, as suggested by Fig. 8.5, these transformations can distort a line into a knotted line since the quality of being a continuous curve with its ends not joined is not destroyed. It is this connectedness, this continuity, of lines and surfaces which is preserved by topological transformations. This means that these transformations do not disconnect what had been connected (i.e., no cutting or puncturing of holes) nor do they connect what had been separated (i.e., no welding, glueing, etc.). The only time tearing or cutting of a surface or curve is permitted is when the surface or curve is rejoined after the cut in such a way that points along the cut "match up" identically to their positions prior to the cutting. For example, a topological transformation may take a simple closed curve and knot it by cutting and rejoining (Fig. 8.6).

When a three-year-old is asked to make a copy of a square or triangle, he may draw both as a simple closed curve, which appears somewhat like a circle. Another child may draw a picture of a person with the eyes outside of the head. As the child approaches age six, these difficulties are corrected and the basic topological ideas of the child are refined and expanded.

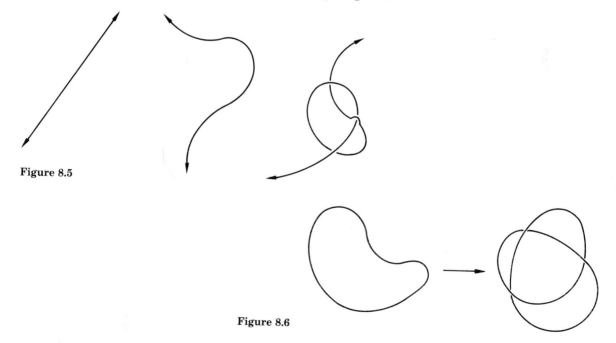

Figure 8.5

Figure 8.6

Some interesting pictures for students regarding topological equivalents appear in the Life Science Library book entitled *Mathematics*. A pictorial description is given showing how to remove one's vest without removing one's coat. This is very easy to do and students enjoy a demonstration.

When one surface or curve can be transformed by a topological transformation to another surface or figure, we say that the two figures are **topologically equivalent**. That is, within the realm of topology two topologically equivalent surfaces are indistingishable—they are topologically alike. Figure 8.1 illustrates that a circle and an equilateral triangle are topologically equivalent. Also, the simple closed curve and the knot of Fig. 8.6 are topologically equivalent. Does Fig. 8.7 convince you that a length of pipe and a washer are equivalent? Finally, in Fig. 8.8 we see how topologists got their reputation. A donut and a coffee cup are topologically equivalent.

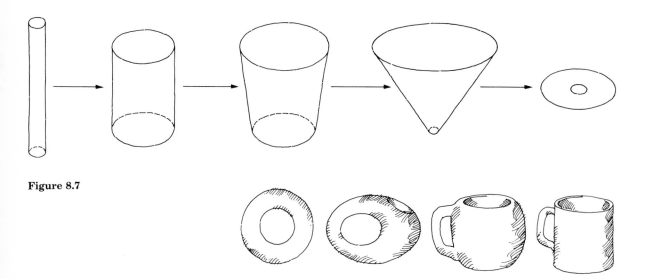

Figure 8.7

Figure 8.8

EXERCISE SET 8.2

1. Which of the figures presented in parts (a), (b), (c), and (d) are topological distortions of the figure on the extreme left? That is, which of the figures are topologically equivalent to the figure on the left?

(a)

(b)

(c)

(d)

2. Which pairs of letters printed below are topologically equivalent?

AEFGLRW

3. Each of the points A, B, C, D below is an example of what is called a **cut point** because if these points are removed, the figure is separated into several (more than two) disjoint sets.

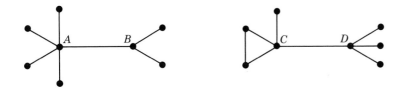

In each of the figures below find all of the cut points.

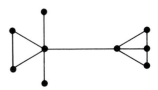

Cut point
of order 3

Cut point
of order 5

4. If a cut point P, when removed from a figure, separates the figure into exactly n disjoint pieces, we call it a **cut point of order n**. For each of the cut points found in the figures in Exercise 3, describe the order of cut point.

5. It is intuitive, and possible to prove, that two geometric figures are not topologically equivalent if they do not have the same number of cut points of each order. Using the "cut point of order n" criteria, select the one figure which is not topologically equivalent to the other two in parts (a), (b), (c), and (d).

(a)

(b)

(c)

(d)

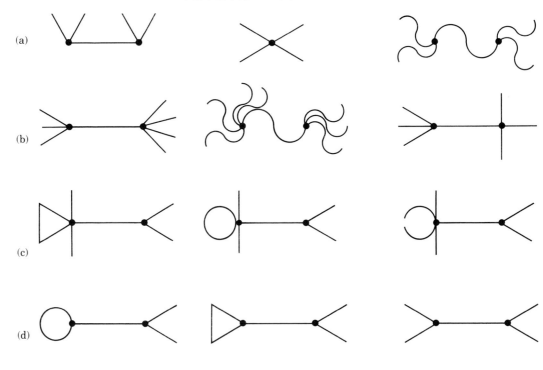

6. In each part below complete the figure on the right so that it is topologically equivalent to the figure on the left.

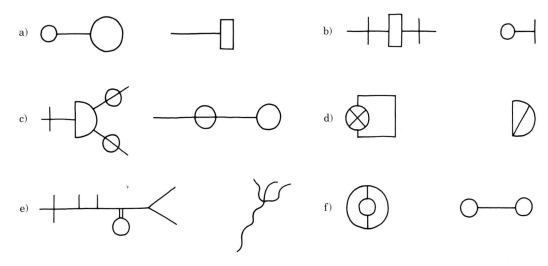

7. How many of the 14 semiregular polyhedra studied in Chapter 4 seem to you to be topologically equivalent to the sphere?

*8. The figure on the right is sometimes seen on the shoulder of a uniform. We shall call it a woggle. The four stages of the figure below demonstrate that a woggle is topologically equivalent to a strip with two slits made in it.

Try to make a woggle.

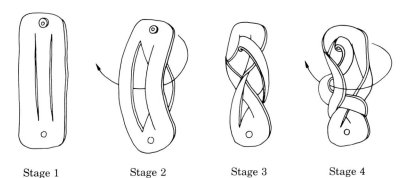

Stage 1   Stage 2   Stage 3   Stage 4

NETWORKS

In the remaining four sections of this chapter we shall shift our attention from topological transformations themselves to properties of curves and surfaces which remain unchanged by topological transformations. The properties of "has no ends," "has exactly one edge," and "has exactly one surface" are all examples of properties which are preserved under topological distortion which we have already encountered in this chapter.

We continue by introducing in the next investigation a famous problem which originated in the city of Kaliningrad.

Investigation 8.2

The city of Konigsberg (now Kaliningrad) in East Prussia is located on the banks and two islands of the Pregel River. The various regions of the city were connected by seven bridges as shown in the figure. A popular question in the early eighteenth century was the following: Is it possible to take a walk around town is such a way that, starting from home, one can return there after having crossed each bridge just once? Explore this question.

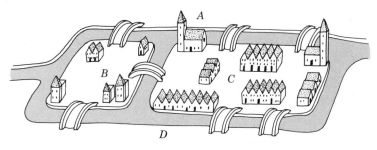

What did you find in your investigation of the Konigsberg bridges? Note that the bridge system can be drawn in a much simplified form by drawing each land mass *A,B,C,D* as a point and each bridge as a curve connecting the appropriate points, resulting in Fig. 8.9. This diagram is an example of what is commonly called a network. Other examples of networks are

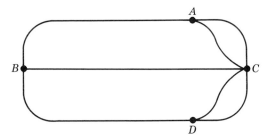

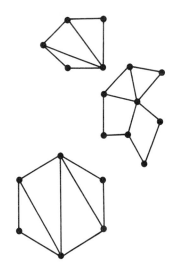

Figure 8.9

shown in Fig. 8.10. A **network** is a collection of points, called **vertices**, together with a collection of curves, called **edges**, which join the vertices.

Before answering the Konigsberg bridge question, we shall detour for a brief study of network theory. In the process of this study we shall find an answer to the Konigsberg problem.

A familiar question asked in network theory is, *Can we trace a path on a given network which traverses each edge once and only once?* For example, for a given network of highways, a highway inspector wants to inspect each of the highways without the wasted effort of traveling any highway more than once. In fact, we see that the Konigsberg bridge question can be restated as: Is it possible to trace a path along the network in Fig. 8.9 which traverses each edge once and only once and which ends at its beginning point? If so, the network is called **traversable**. If the beginning point and ending point of such a path coincide, the network is called **traversable type 1**; if the beginning point and ending point do not coincide, the network is called **traversable type 2**. In Fig. 8.11 we see an example of a type 1 network, a type 2 network, and a network which is not traversable without an edge being covered twice or more.

Figure 8.10

Figure 8.11

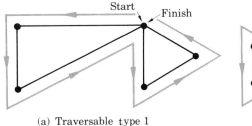

(a) Traversable type 1

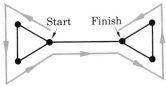

(b) Traversable type 2

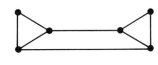

(c) Not traversable

Investigation 8.3

A. Complete the table for networks 1 through 10 below. Note
that some may not be traversable either type 1 or type 2.

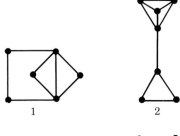

1

2

3

4

Network	Number of even vertices	Number of odd vertices	Traversable type 1 (yes or no)	Traversable type 2 (yes or no)
1				
2				
3				
4				
5				
6				
7				
8				
9				
10				

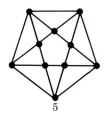

5

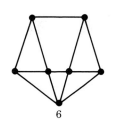

6

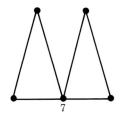

7

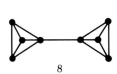

8

9

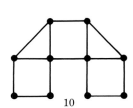

10

B. Reflect on the table in part A. What generalizations and
conjectures are suggested by the relationships expressed
in this table?

In the Konigsberg network in Fig. 8.9 there is an odd number of edges joined at vertex A, and so A is called an **odd vertex.** There are five edges joined at vertex C so C is also an odd vertex. When there is an even number of edges joined at a vertex, it is called an **even vertex.** The number of even and odd vertices to a network influences whether or not it is traversable. See if you can discover in Investigation 8.3 a rule which determines when a network is traversable type 1 and when it is traversable type 2. Apply the rule that you discover to answer the Konigsberg bridge question.

On the basis of the experience gained in completing this investigation, which of the following conjectures do you believe are true?

Conjecture 1. If a network has all of its vertices of even order, then the network is traversable type 1 and type 2. Furthermore, a traversable path may begin at any vertex.

Conjecture 2. If a vertex of a network is an odd vertex, then a traversable path (if it exists) must either begin or end at that vertex.

Conjecture 3. If in a network exactly two vertices are odd vertices, then the network is traversable type 2. Any traversable path must begin at one of these vertices and end at the other vertex.

Conjecture 4. If a network has two or more vertices which are odd vertices, then it is not traversable type 2.

Some of the ideas on networks presented here can often be couched in more popular language. For example, a student can be asked, "Can you draw this figure without lifting your pencil from the paper?" The conclusions from Investigation 8.3 should give an immediate method for deciding on answers to questions like this.

EXERCISE SET 8.3

1. Use the network in Fig. 8.9 and explain why a walk cannot be taken in Konigsberg so that each of the bridges is crossed exactly once and the walker returns to the starting place.

2. A city planner commented that any one of the seven bridges at Konigsberg could be relocated so that a walk could be taken crossing each bridge exactly once.

   Show that the city planner was correct by drawing networks representing the bridges before and after relocation.

3. The networks below are not traversable. What is the fewest number of edges which, when added, makes the network traversable type 2? traversable type 1?

  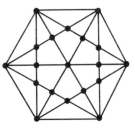

4. Can you draw a continuous path crossing each edge of the figures on the left just once without going through a corner?

   a) Seek an answer by trial and error.

   b) Seek an answer by translating this question into a consideration of the traversability of a related network. [*Hint*: Let each region become a vertex and each boundary edge between regions become an edge.]

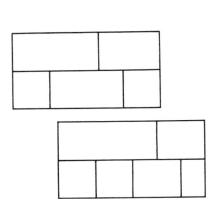

5. Can a person find a continuous path in this house that will take him through each door exactly once?

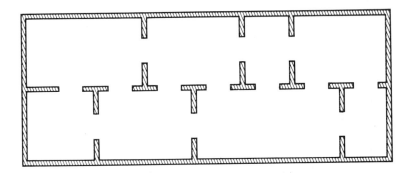

6. A network and its distorted image under a topological transformation have some properties in common. That is, some properties of networks are preserved by topological transformations. List some of these properties.

JORDAN CURVE THEOREM

The Jordan Curve Theorem concerns simple closed curves in the plane and is one of the most famous theorems in topology. Although we shall not prove this theorem, we shall experience it on the intuitive level. We begin with an investigation.

Investigation 8.4

Given three houses A, B, and C and the utility centers of electricity (E), gas (G), and water (W), connect each of the three houses to each of the three utilities without the lines crossing.

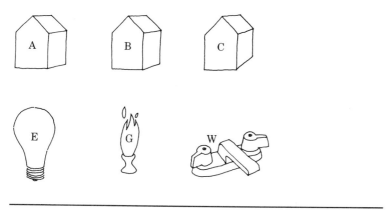

Are you convinced that you have found a solution to the utilities problem? We shall base our analysis of the problem on the concept of a simple closed curve. A **simple closed curve** is a curve which is the image of a circle under a topological transformation. In Fig. 8.12 there are four curves in the plane. Only curves (a), (b), and (d) are simple closed curves.

Suppose that in searching for a solution to the utilities problem we proceed by joining A with each utility and B with each utility as indicated in Fig. 8.13. This process yields a simple closed curve $AWBGA$ with E on the "inside" and C on the "outside" of this curve. Any path which joins E to C would

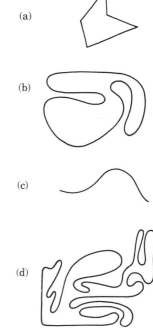

(a)

(b)

(c)

(d)

Figure 8.12

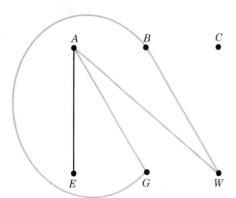

A   B   C

E   G   W   **Figure 8.13**

Since the notions of inside and outside are available to very young children, ideas of simple closed curves can be presented quite early. Young children like to solve simple mazes involving closed curves or open curves. Older students enjoy inventing mazes. Some challenging mazes occur in the book *Mazes* (Grosset and Dunlap, 1971) by Vladimir Koziakin.

then cross the simple closed curve *AWBGA*, and hence it would cross one of the utility connections. The reader should convince himself that any attempt at a solution to this problem results in a simple closed curve with a point on its inside and a point on its outside.

We have in fact been assuming in this explanation of the utilities problem the famous Jordan Curve Theorem — a theorem easily stated, readily accepted, but surprisingly difficult to prove. The essence of Jordan's theorem is that a simple closed curve separates the plane into two subsets — one we call "inside", the other "outside" — in such a way that each continuous path from any point inside the curve to any point outside the curve crosses the curve.

It is sometimes surprising how difficult it is to determine whether a given point is in what we commonly call the inside or on the outside of of a simple closed curve.

Figure 8.14

For example, in Fig. 8.14 is point *A* on the inside or the outside? This type of question is the subject of the next investigation.

Investigation 8.5

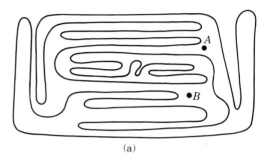

(a)

A. Figure (a) shows a simple closed curve and two points *A* and *B*. Which point is outside the curve? Can you reach *B* from *A* without crossing the curve?

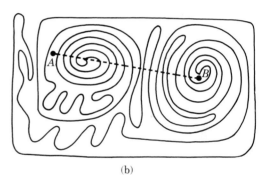

(b)

B. Figures (b) and (c) each show a simple closed curve and two points *A* and *B*. Is *A* inside the curve in each case? Are *A* and *B* one inside and one outside the curve? [*Hint:* How many times does the dashed line cross the curve?]

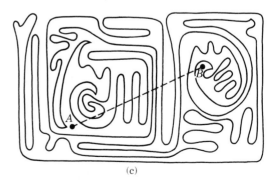

C. Draw some of your simple closed curves. Find a rule for deciding which points are inside your curves.

(c)

CONTIGUOUS REGIONS

In this section we shall study relationships between a region in the plane and the regions surrounding it. We begin with an investigation.

Investigation 8.6

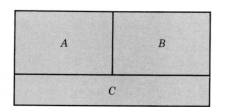

Consider the following story. A man with three sons stipulated in his will that his land was to be divided among his sons so that every son would be a neighbor to each of his brothers. (Being neighbors means that their land touches along a boundary line, rather than only at a point.) That is, each pair of regions must be contiguous along a boundary line. For example, with three sons the following division satisfies this condition.

Can you find such a division for four sons? five sons?

To analyze this problem we first observe that since the relation "neighboring regions" remains unchanged through topological distortion, it is sufficient to draw our regions as rectangular.

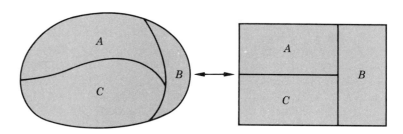

We analyze the problem by finding solutions for the case that there are two sons and hence only two regions. Then we solve the problem in the case where there are three regions by considering all the ways one can add a third region to each solution in the case of two regions. We continue this process until we arrive at the case of 5 regions.

If there are only two sons, there are only two essentially different divisions for the land. In one of the divisions one region completely surrounds the other.

If there are three sons there are three possible divisions, one in which two regions are completely surrounded, one with one region completely surrounded, and one with no regions completely surrounded. These three regions correspond to three different ways of adding a third region to Fig 8.15 (b). Note that any third region added around the outside of Fig. 8.15 (a) would not be a neighbor of the inner region.

In Fig. 8.16 we see that when any one region becomes completely surrounded, the process of adding regions which

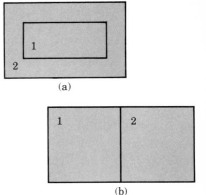

(a)

(b)

Figure 8.15

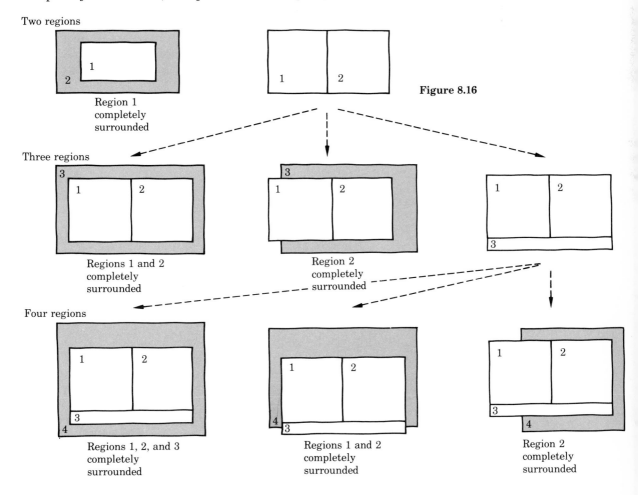

Two regions

Region 1 completely surrounded

Figure 8.16

Three regions

Regions 1 and 2 completely surrounded

Region 2 completely surrounded

Four regions

Regions 1, 2, and 3 completely surrounded

Regions 1 and 2 completely surrounded

Region 2 completely surrounded

are neighboring to all others must stop. Furthermore, in order for the fourth region to be neighboring to each of the first three regions, it must completely surround one region. Consequently, it appears that it is impossible to divide land into five regions with each region neighboring the remaining four.

FOUR-COLOR PROBLEM

Students at various levels enjoy exploring the coloring problem. Young children might, for example, attempt to color a portion of a map of the United States using as few colors as possible according to the condition that neighboring regions are colored differently. Older students will be interested in the problem and the fact that the Four Color Conjecture, while easily understood, has been neither proven nor disproven.

A property of maps observed empirically by English cartographers was that four colors was sufficient to color a map so that neighboring regions (i.e., regions with a common boundary) are colored differently. (Two regions which touch only at a vertex may be colored alike.) In 1850 Francis Guthrie, a student of mathematics at Edinburgh, noted that if this observation were always true, it would be an interesting mathematical theorem. Guthrie's conjecture was that in an arbitrary geographical map where regions represent a political subdivision, four or fewer colors are always sufficient to color the map so that any two regions touching along a boundary line have different colors. In 1878 Arthur Cayley communicated this conjecture to the London Mathematical Society. To this day this conjecture has neither been proven nor disproven.

This so-called "Four-color problem" is of course related to the contiguous region problem of the last section. For certainly, a map containing four contiguous regions would require four colors—one for each region.

In Investigation 8.7 we explore some of the relationships between the contiguous region and four-color problems.

Several observations should be made about this investigation. Part (c) is a map with at most two neighboring regions and which requires three colors. Part (e) is a map with at most three neighboring regions and which requires four colors. So we see that in some cases one more color is required than the number of neighboring regions. Since it is possible to have a map with four neighboring regions, as we have seen in Fig. 8.16, doesn't it seem plausible that there might exist a map which requires more than four colors? The four-color conjecture claims that the answer is no. Do you believe it?

Investigation 8.7

For each of the maps below find:

a) the minimum number of colors required to color the map so that neighboring regions are of a different color;

b) the maximum number of regions all neighboring to one another.

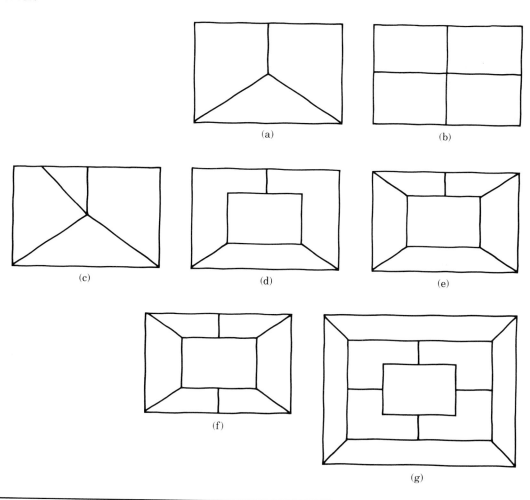

(a)

(b)

(c)

(d)

(e)

(f)

(g)

EXERCISE SET 8.4

Trace and color this section of a map of the United States so that neighboring states do not have the same color. How many colors are required?

BIBLIOGRAPHIC REFERENCES

[6, 26–33, 50–61, 204–215], [8], [10], [15, 176–190], [17, 39–44], [20, 12–20], [24, 235–270], [25], [29, 191–195], [38], [40, 9–18, 91–96], [41, 63–72], [50, 290–333], [61], [73], [80], [83, 13–17, 61–66, 73–82], [99, 64–83, 226, 236]

PEDAGOGICAL ACTIVITIES FOR THE TEACHER

1. Engage students at a given level in an experiment with the Mobius Strip. Note their reactions.

   a) Did the students get involved and interested in the experiment?

   b) Did the students make any conjectures? That is, did they say, "I wonder if this might be true?"

   c) Did the students ask any questions about properties of the strip other than questions on mechanical procedure?

   d) Did the students have a tendency to want to carry the experiment further, or were they satisfied with the initial experience?

   e) How would you change the experiment to make it more challenging or interesting?

2. At what age can students understand topological equivalence? Investigate this question by devising a mini-lesson which would develop this concept. Include some exercises which you could present to students to see if they have understood the idea. You may want to present this mini-lesson to individual students at different grade levels.

3. Consult books such as *The Scientific American Book of Mathematical Puzzles and Diversions* (Simon and Schuster, 1959) and *New Mathematical Diversions* from *Scientific American* (Simon and Schuster, 1956) and collect at least five topologically oriented puzzles which you feel would be appropriate for students at a given grade level. Include these puzzles in your card file of mathematical recreations.

4. Create a sequence of concept cards which use examples and nonexamples to develop the idea of topological equivalence. Use the cards with students and evaluate their effectiveness. Revise the cards for classroom use.

5. Study Chapters 8 and 9 of Copeland's book, *How Children Learn Mathematics* (Macmillan, 1970). On the basis of this study, create some activities which you could use with preschool and primary school children to check their developmental stage with regard to topological notions. Try these activities with children and summarize your findings.

9

Number Patterns in Geometry

INTRODUCTION

The important concept of "pattern" is so broad that any attempt to define it will likely restrict its meaning. And yet, nearly everyone has an intuitive understanding of the concept. In this chapter we think of pattern in the sense of being a reoccurring configuration or relationship. W. W. Sawyer has said:

> Mathematics is the classification and study of all possible patterns . . . In mathematics, if a pattern occurs, we go on to ask, Why does it occur? What does it signify? . . .*

In order to restrict this rich subject to a "bite-size portion," we shall limit our discussion in this chapter primarily to certain patterns of points and lines and the number patterns which emerge from these geometric settings.

PATTERNS OF POINTS

Thinking of point in the physical sense, we observe that there are both natural and man-made settings in which patterns of points occur.

We see this in X-ray diffraction patterns (Fig. 9.1), in a magnification of a crystal of virus (Fig. 9.2) or in a photograph of portions of the solar system (Fig. 9.3).

The computer printout of prime numbers (Fig. 9.4) represents man-made patterns of points.

Some of these patterns may be useful to the scientist for laboratory analysis while others are simply aesthetically pleasing. Regardless of our purpose, if we are sensitive to patterns in the world around us, even the physical notion of point can take on new meaning. Investigation 9.1 will provide an opportunity for a closer look at some simple patterns of points.

---

* From W. W. Sawyer, *Prelude to Mathematics*. (Baltimore: Penguin Books, 1955.)

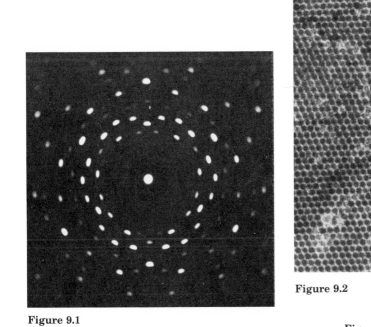

Figure 9.1

Figure 9.2

Figure 9.3

				100
				200
				300
				400
				500
				600
				700
				800
				900
				1000

Figure 9.4

Investigation 9.1

A geoboard is a board with a square array of nails (points) on it. Here are a few different-sized geoboards.

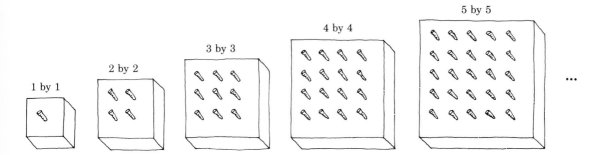

If you place rubber bands on these geoboards in special ways and count the nails inside them, you produce a sequence of numbers. Write the first eight numbers in the sequence produced in each situation presented here.

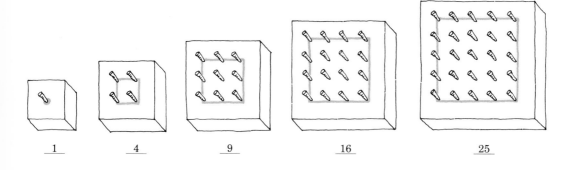

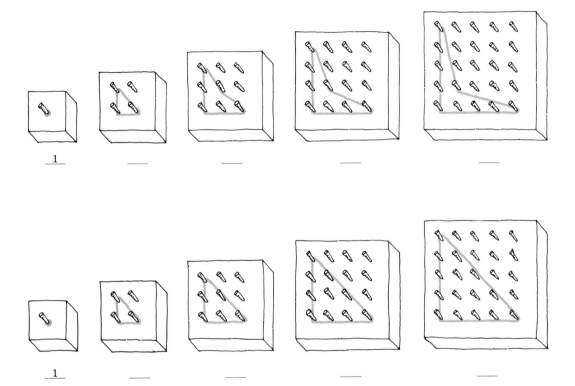

1 ____   ____   ____   ____   ____

1 ____   ____   ____   ____   ____

For each situation, can you figure out how to find the number of nails inside the rubber band on a 100 by 100 geoboard? on an *n* by *n* geoboard? (Write a formula.)

Now make some more "geoboard sequences" of your own and answer the same questions about them.

This investigation of simple "point patterns" on geoboards demonstrates that number generalizations can be illustrated and clarified by viewing geometric patterns. For example, the pattern produced by enclosing all nails on each geoboard suggests the sequence of numbers called **square numbers**. Table 9.1 describes the situation.

Geometric Situation			Numerical Situation	
Size of geoboard	Total number of nails enclosed		n	The nth square number is
1 by 1	•	1	1	1
2 by 2	∶∶	4	2	4
3 by 3	⋮⋮⋮	9	3	9
.		.	.	.
.		.	.	.
.		.	.	.
n by n		?	n	?

Table 9.1

By observing the geometric situation we find that the generalization "the nth square number is n^2" takes on a fuller meaning.

Similarly, the pattern produced by enclosing the triangular array of nails on the geoboard suggests the sequence of numbers called **triangular numbers.** Table 9.2 describes the situation. (Note that the black dots indicate the nails that were enclosed on the geoboard. The gray dots have been included to display rectangular arrays of dots which are helpful in forming the generalization. Observe that each triangular array contains half the number of dots in the rectangular array.)

By observing the arrays of dots we see that the number of dots enclosed on a geoboard with 2 nails on a side is $\dfrac{2 \times 3}{2}$ (i.e., width of array × length of array ÷ 2), the number on a

If the teacher sees mathematical patterns as a source of beauty and interest, then the student may also come to recognize this beauty and look for it in a new situation. The geoboard is a valuable device for creating situations where children can discover patterns. The teacher might be interested in Dan Cohen's book *One Hundred and Fifty Activities for the Geoboard,* available from Walker and Walker.

Geometric Situation			Numerical Situation	
Size of geoboard	Total number of nails enclosed		n	nth triangular number
1 by 1	•	1	1	1
2 by 2		3	2	3
3 by 3		6	3	6
4 by 4		10	4	10
·		·	·	·
·		·	·	·
·		·	·	·
n by n		?	n	?

Table 9.2

geoboard with 3 nails on a side is $\dfrac{3 \times 4}{2}$, and the number on a geoboard with 4 nails on a side is $\dfrac{4 \times 5}{2}$. This geometric discovery suggests the numerical generalization that "the nth triangular number is $\dfrac{n(n + 1)}{2}$."

In the situations above we associated numbers with a geometric situation. This practice is common throughout the study of geometry and results in generalizations involving both numbers and geometric content. Generalizations are valuable in all areas of mathematics because they allow us to economize our effort. Rather than dealing with each of several specific cases as if they were different, we form a generalization which gives us a way of dealing with any of the several cases in the same way.

It is important, then, to gain experience in recognizing patterns and finding generalizations, so we continue exploring these patterns. For example, consider the display in Table 9.3, which involves another way of looking at square arrays of dots.

Searching for patterns might well be the most important experience the student will have in mathematics. These experiences develop habits and skills of pattern seeking that can be extremely valuable as the student encounters new ideas. The book *Notes on Mathematics for Primary Schools* (Cambridge University Press) suggests several geometric situations rich in patterns.

Dot pattern	Numerical relationship
•	$1 = 1^2$
	2 addends
	$1 + 3 = 2^2$
	3 addends
	$1 + 3 + 5 = 3^2$
	4 addends
	$1 + 3 + 5 + 7 = 4^2$

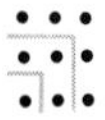

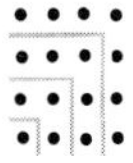

Table 9.3

By studying the dot pattern, the generalization "the sum of the first n odd numbers is n^2" becomes clear. In this case, an interesting way of viewing a square array of dots suggests a valuable shortcut for finding the sum of consecutive odd integers, starting with 1. Another similar example involves triangular arrays of dots. See Table 9.4.

Dot pattern	Numerical relationship
•	$1 = \dfrac{1 \times 2}{2} = 1$
	$1 + 2 = \dfrac{2 \times 3}{2} = 3$
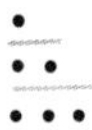	$1 + 2 + 3 = \dfrac{3 \times 4}{2} = 6$
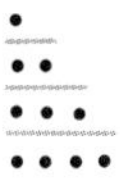	$1 + 2 + 3 + 4 = \dfrac{4 \times 5}{2} = 10$

Table 9.4

The reader should study the patterns above and complete the generalization: "The sum of the first n whole numbers is ———."

Whether in the context of simple patterns of dots or of complicated patterns of points, lines, and other geometric figures, the search for generalizations in geometry continues. Whenever these geometric generalizations can clarify or suggest numerical generalizations, so much the better. Whenever a numerical generalization can shed light on our understanding of geometry, we experience the value of the close relationship between number and space.

Patterns often can be described using mathematical functions. A function machine has an input, operates according to a rule, and produces an output. A record of the machine's operation can be kept in a table where the input and output are listed as ordered pairs. A function machine such as the one shown below can be used to help students discover relationships utilizing geometric figures as well as numbers.

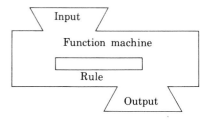

EXERCISE SET 9.1

1. a) What is the 50th square number? the 100th square number?

   b) What is the 50th triangular number? the 100th triangular number?

2. a) What is the sum of the first 50 odd numbers? the first 100 odd numbers?

   b) What is the sum of the first 50 whole numbers? the first 100 whole numbers?

3. A rectangular number is a number which, like 6, can be illustrated with a rectangular array of dots. The number 5 is *not* a rectangular number, since neither a single row of dots nor the array with a single extra dot is considered a rectangular array.

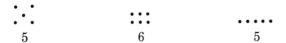

   a) Write the numbers up to 35 that are rectangular numbers.

   b) Use the idea of "square" numbers and "rectangular" numbers to write a definition of prime numbers.

4. If you start with the square arrays of dots and draw lines like this, it seems natural to write these equations:

$$1 = 1^2 \qquad 1 + 2 + 1 = 2^2 \qquad 1 + 2 + 3 + 2 + 1 = 3^2$$

Write the next three equations in the sequence.

State a generalization about this relationship.

5. If you start with square arrays of dots and draw lines like this, a pattern emerges.

$$1 + 3 = 2^2 \qquad 3 + 6 = 3^2 \qquad 6 + 10 = 4^2$$

Write the next three equations.

State a generalization about this relationship. [*Hint*: Note that 1, 3, 6, 10, etc. are triangular numbers.]

6. Suppose you enclose nails on geoboards like this and count all nails enclosed on each board.

   a) Write the first eight terms in the sequence of numbers determined in this situation.

   b) How many nails are enclosed on a 100 by 100 geoboard? an n by n geoboard?

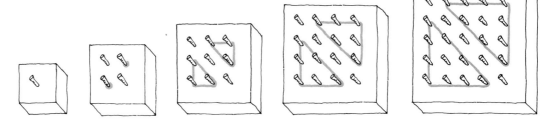

7. Pentagonal numbers can be described in a manner similar to that used to describe triangular and square numbers. Here are the first four pentagonal numbers.

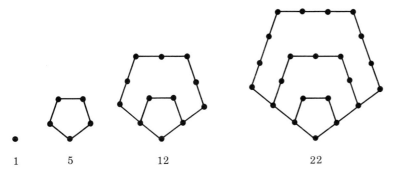

| 1 | 5 | 12 | 22 |

What are the next three pentagonal numbers? Show the dot patterns for these numbers.

8. Use the idea suggested in Exercise 7 to find the first seven hexagonal numbers and show their dot patterns.

*9. Express each pentagonal number and hexagonal number as a sum in such a way that a "pattern of sums" emerges similar to those we observed for triangular and square numbers.

*10. We have seen ways to geometrically express "triangular" and "square numbers." How would you geometrically express a "cube number"?

*11. a) Study the following explanation:

Algebraic methods can be used to find the formula for a relationship shown by a table. We illustrate such a method using the table for triangular numbers shown at the right.

Since the second difference is constant (always equal to the same number), it can be shown that the equation of $T(n)$ is a quadratic equation of the form

$$T(n) = an^2 + bn + c.$$

We proceed to solve simultaneous equations to find values for a, b, and c.

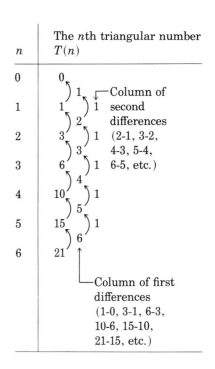

The nth triangular number

n	$T(n)$
0	0
1	1
2	3
3	6
4	10
5	15
6	21

Column of second differences (2-1, 3-2, 4-3, 5-4, 6-5, etc.)

Column of first differences (1-0, 3-1, 6-3, 10-6, 15-10, 21-15, etc.)

Note that:

$$T(0) = a \cdot 0^2 + b \cdot 0 + c = 0$$
$$T(1) = a \cdot 1^2 + b \cdot 1 + c = 1$$
$$T(2) = a \cdot 2^2 + b \cdot 2 + c = 3$$

Simplifying, we have the equations

$$c = 0,\ a + b = 1,\ \text{and}\ 4a + 2b = 3.$$

Solving these simultaneously, we have

$$a = \tfrac{1}{2},\ b = \tfrac{1}{2},\ \text{and}\ c = 0.$$

n	$P(n)$
0	0
1	1
2	5
3	12
4	22
5	35
6	51

Hence $T(n) = \tfrac{1}{2}n^2 + \tfrac{1}{2}n + 0$ or $T(n) = \dfrac{n^2 + n}{2}$.

b) Use the method explained in (a) and this table of pentagonal numbers to find the equation which can be used to find the nth pentagonal number, $P(n)$.

12. If n represents the number of dots in a row and $P(n)$ represents the number of different pairs of dots, construct a table showing the relationship between n and $P(n)$.

*13. Complete the table and determine the pattern.

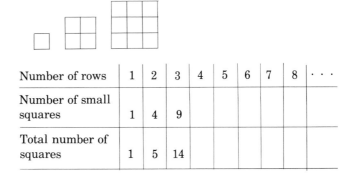

Number of rows	1	2	3	4	5	6	7	8	$\cdots$
Number of small squares	1	4	9						
Total number of squares	1	5	14						

n	$P(n)$
0	0
1	1
2	5
3	12
4	22
5	35
6	51

*14. The numbers inside the loop in this table of pentagonal numbers can be used in a short cut method for finding the equation which can be used to find the nth pentagonal number.

Using the equation

$$P(n) = an^2 + bn + c,$$

$a = $ (constant second difference) $\div 2 = {}^3\!/_2$
$b = $ (1st first difference) $- a = 1 - {}^3\!/_2 = -{}^1\!/_2$
$c = P(0) = 0.$

Hence the equation is

$$P(n) = \frac{3}{2}n^2 + \frac{-1}{2}n + 0.$$

The tables below express numerical relationships that might describe some geometric situation. Use the shortcut explained above to write an equation for each which can be used to find the $f(n)$ entry in the table for any value of n.

n	$f(n)$
0	3
1	4
2	9
3	18
4	31
5	48
6	69

a)

n	$f(n)$
0	2
1	5
2	9
3	14
4	20
5	27
6	35

b)

n	$f(n)$
0	1
1	5
2	11
3	19
4	29
5	41
6	55

c)

*15. a) Figure out a way to use the rows of triangular and square numbers to find the pentagonal numbers. How did you do it?

b) Use the rows of square and pentagonal numbers to complete the row of hexagonal numbers.

Triangular	1	3	6	10	15	21	28	36	45	55	66
Square	1	4	9	16	25	36	49	64	81	100	121
Pentagonal	1	5	12								
Hexagonal	1										

*16. Complete the table below by observing number and formula patterns.

Polygonal numbers	First	Second	Third	Fourth	Fifth	$\cdots$	nth
Triangular	1	3	6	10	15	$\cdots$	$\dfrac{n(1n - {}^-1)}{2}$
Square	1	4	9	16	25	$\cdots$	$\dfrac{n(2n - 0)}{2}$
Pentagonal	1	5	12	22	35	$\cdots$	$\dfrac{n(3n - 1)}{2}$
Hexagonal	1	6	15	28	45	$\cdots$	$\dfrac{n(4n - 2)}{2}$
Heptagonal	1	7	18			$\cdots$	
Octagonal	1	8	21			$\cdots$	
Enneagonal	1	9	24			$\cdots$	
Decagonal	1	10	27			$\cdots$	

PATTERNS OF LINES

Patterns of lines often appear in interesting and exciting ways in everyday situations. Photographs from the world of art and architecture provide illustrations of interesting line patterns.

Many of the familiar mathematical curves and surfaces can be demonstrated with patterns (formally called envelopes) of lines. On the facing page we see examples of the parabola, cardioid, and hyperbolic paraboloid generated in this way. Patterns of lines of this type are often constructed by students involved in curve-stitching activities.

The following investigation provides an opportunity for you to explore line patterns and search for generalizations.

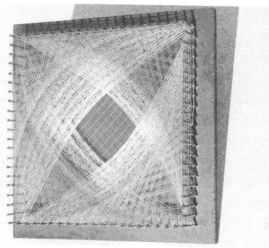

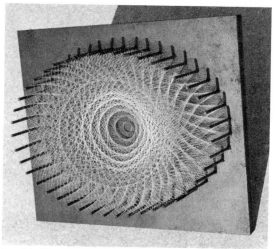

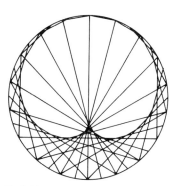

Cardioid

Hyperbolic paraboloid from *Mathematical Snapshots* by H. Steinhaus, 3rd edition. Copyright © 1950, 1960, 1968, 1969 by Oxford University Press, Inc. Reprinted by permission.

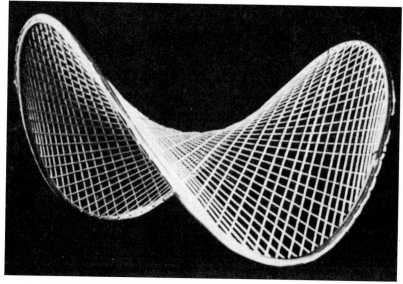

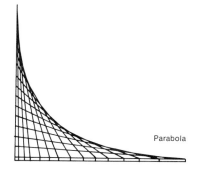

Parabola

Investigation 9.2

Can you draw pictures to help you complete five rows of each table and write an equation expressing the relationship involved?

A.

Number of points marked on a line	Number of different segments that can be named
2	1 ($\overline{AB}$)
3	3 ($\overline{AB}, \overline{BC}, \overline{AC}$)
4	
5	
6	
.	
.	
.	
n	?

B.

Number of points (no 3 on the same line)	Number of different lines determined by these points
2	1
3	3
4	
5	
6	
.	
.	
.	
n	?

C.

Number of lines	Maximum number of intersections of these lines
2	1
3	3
4	
5	
6	
.	
.	
.	
n	?

Investigation 9.2 illustrates that a single numerical relationship can describe several seemingly unrelated geometric
situations. Also, once a relationship such as that for finding
the nth triangular number, $T_n = n(n + 1)/2$, is known, it can
be useful in finding other relationships. For example, consider
the following situation and the table of number pairs below
which describe the relationship.

i) One line divides the plane into two regions.

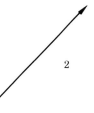

ii) Two lines divide the plane into four regions.

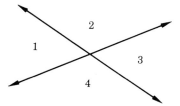

iii) Three lines divide the plane into a maximum of seven
regions.

Number of lines (n)	Maximum number of regions (R_n)
1	2
2	4
3	7
4	11
5	16
.	
.	
.	
n	?

iv) Four lines divide the plane into a maximum of 11 regions
and five lines divide the plane into a maximum of 16
regions (the reader should make drawings to show that
this is true).

When the table for this relationship is compared with the table for the nth triangular number, we have these results.

n	T_n	n	R_n
1	1	1	2
2	3	2	4
3	6	3	7
4	10	4	11
5	15	5	16
.	.	.	.
.	.	.	.
.	.	.	.
n	$\dfrac{n(n+1)}{2}$	n	?

Note that the maximum number of regions, R_n, is one more than the triangular number for a given value of n. Thus the formula for the maximum number of regions is found as follows:

$$R_n = \frac{n(n+1)}{2} + 1 = \frac{n(n+1)}{2} + \frac{2}{2}$$

$$= \frac{n(n+1)+2}{2} = \frac{n^2+n+2}{2}$$

The following exercises present interesting line patterns.

EXERCISE SET 9.2

1. If you draw closed figures with 3, 4, and 5 segments as sides, you can draw 0, 1, and 2 diagonals from a selected vertex.

Number of sides	Number of diagonals from one vertex
3	0
4	1
5	2
6	
7	
8	
.	
.	
.	
n	?

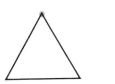 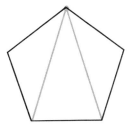

Complete three more rows of this table. What patterns do you see? Write an equation to show this relationship.

2. A triangle has no diagonals. A square has a total of two diagonals.

a) How many diagonals does a five-sided figure have?

b) Complete four more rows of this table. What patterns can you find in this relationship?

Number of sides	Number of diagonals
3	0
4	2
5	
6	
7	
8	
.	
.	
.	
n	?

3. If curved lines are drawn and it is agreed that *every line shall cut every other line twice*, we get these pictures and this table:

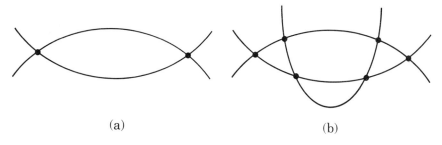

Number of lines	Number of crossings
2	2
3	6

(a)   (b)

Continue the drawings to find three more entries in the table. What patterns can you find in this relationship?

4. A region of the plane that is bounded on all sides by a line is called a **bounded region**.

When three lines intersect, one bounded region is formed.

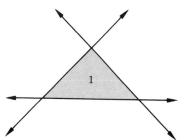

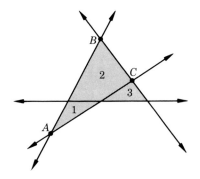

When four lines intersect, three bounded regions are formed. (Note that a region like *ABC* is not counted since it overlaps with regions 1 and 2.)

a) Complete the next two rows of the table.

b) Write an equation, if possible, to describe this relationship.

Number of lines	Maximum number of bounded regions
3	1
4	3
5	
6	
.	
.	
.	
n	?

*5. Instead of counting the total number of regions as in Exercise 4, count the maximum number of *triangular* regions.

Four lines determine two triangular regions (△*BEF* and △*CFG*).

We do not count △*ABC* nor △*AEG* since they overlap with △*BEF* and △*CFG*.

Number of lines	Maximum number of triangular regions
3	1
4	2
5	5
6	7
7	11
8	15
9	21
.	.
.	.
.	.
n	?

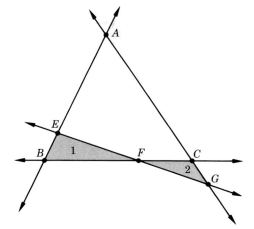

Kobon Fujimura, a Japanese puzzle expert, posed this problem. The entries in the table are thought, *but not proved*, to be the maximal solutions. At the date of this printing no equation describing this relationship for all *n* has yet been found.

Can you draw diagrams for each number of lines to show the number of regions indicated in the table?

(See issues of *The Scientific American*, April, 1972; and May, 1972 for further information on this problem.)

*6. We have seen that geometrical patterns often suggest numerical patterns. Similarly, number patterns often suggest geometrical patterns. Suppose 36 points evenly spaced around a circle (and numbered 1 through 36) are joined by the pattern 1 to 2, 2 to 4, 3 to 6, . . . , and *x* to 2*x*, remembering that in this context 37 is equivalent to 1, 38 to 2, and so on.

This pattern of lines suggests a curve known as an epicycloid. Make a second and third drawing and explore what curve results if

a) *x* is joined to 3*x*

and

b) if *x* is joined to 4*x*.

When these points are joined by yarn, this process is known as curve stitching.

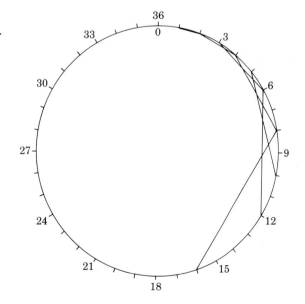

*7. Make five rows of a table showing the maximum number of quadrilateral regions formed by intersecting lines in the plane. (See Exercise 5.)

PATTERN AND PROOF

The process of inferring how a pattern continues from the consideration of a few examples is called **induction**, or sometimes **inductive reasoning**. The following investigation provides an interesting setting for considering this process.

Investigation 9.3

A segment connecting two points on a circle is called a **chord** of the circle. Two points on a circle determine a chord which divides the interior of the circle into two regions. Beginning with three points, three chords and four regions are determined.

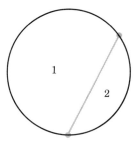
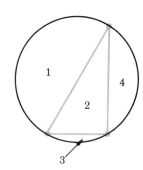

Number of points	Number of regions
2	2
3	4
4	
5	

A. Continue this procedure by marking the specified number of points (unevenly spaced) on the circle, constructing all chords determined by the points, and complete the table at the left.

B. Guess the number of regions generated when beginning with 6, 7, and 8 points.

C. Check the guesses made in part B by going through the construction in each case.

The lesson we learn from this investigation is that we must proceed with a degree of caution when seeking patterns by inductive processes. What we may think at first glance is an obvious pattern may turn out to be a quite subtle one as illustrated in the investigation. This pattern is referred to as the pattern of the "Lost Region." It illustrates that *we cannot be certain that our inductive conclusion about a particular pattern is correct unless it has been proven*. The process of proof is called a **deductive** process, or **deductive reasoning**. We must recognize the importance of the deductive process, based on accepted assumptions and utilizing a system of logic, as an ultimate means of verifying our conjectured generalizations. While this process has not been emphasized in this book, the reader is referred to bibliography references ([4], [24], [36], [68]) as excellent resources for further exploration of geometry as a deductive mathematical system.

Even though the potential for incorrect inferences exists, it is desirable that we explore and make inductive conjectures about possible patterns and relationships. All mathematicians engage in this type of inductive process, for it is an integral part of the development of an intuition about mathematical ideas and a method for discovering possible generalizations to be proven. We must "live dangerously" and be willing to risk making mistakes as we investigate new mathematical situations.

EXERCISE SET 9.3

1. Draw chords of a circle, crossing whenever possible. Then count the number of regions into which these chords divide the circle.

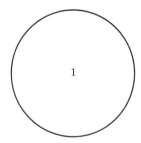

0 chords divide the circle into one region

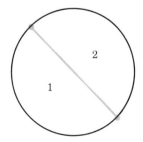

1 chord forms 2 regions

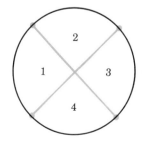

2 chords form 4 regions

Number of segments	Number of regions
0	1
1	2
2	4
.	.
.	.
.	.

Copy the table, *guess* the pattern, and complete three more rows.

Draw circles to check your guess. Were you correct?

2. If we use rubber bands to represent segments, a geoboard with two nails on each side has segments of two different lengths.

A geoboard with three nails on each side has segments of five different lengths.

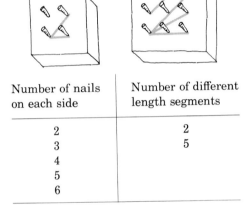

Number of nails on each side	Number of different length segments
2	2
3	5
4	
5	
6	

a) Use a geoboard or dot paper and complete the next two rows of the table.

b) Study the pattern and *guess* the numbers in the next row of the table.

c) Check your guess by drawing the different length segments on 6 by 6 dot paper. (Be careful not to repeat same length segments in different positions. For example, recall that the hypotenuse of a triangle with legs 3 and 4 units long is 5 units long.)

Was your guess correct?

*3. Explore the pattern of differences (see Exercise 11, in Exercise Set 9.1) for the number pattern discovered in Investigation 9.3.

Use the pattern of differences to conjecture about the number of regions generated by 9 and 10 points.

PASCAL'S TRIANGLE AND RELATED PATTERNS

A **polygonal path** is the union of a set of line segments, $\overline{AB}$, $\overline{BC}$, $\overline{CD}$, $\overline{DE}$, etc. with common endpoints.

The following investigation involves a search for the number of different possible polygonal paths from one point to another on the geoboard. For this investigation we shall agree to consider only those paths that always move either "up" or "to the right." In Fig. 9.5, neither of these paths from A to B would be counted.

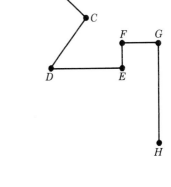

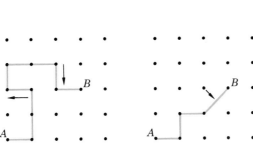

Figure 9.5

This path goes "to the left" and "down"

This path goes along a diagonal

Investigation 9.4

For each nail on the geoboard count the number of paths from A to that nail. Record the number above the nail as is illustrated below. If you can discover some useful number patterns as you begin to complete the array, your work will become much easier. Be careful!

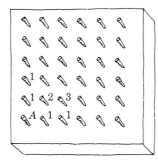

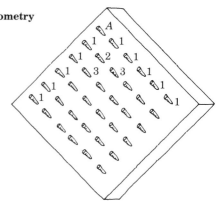

Older students will enjoy a follow up of the idea of Pascal's triangle and the related Fibonacci numbers. These ideas are beautifully described in Chapters 4, 10, 11, and 12 of the book *The Divine Proportion* (Dover Publication, 1970) by H. E. Huntley.

If the geoboard in Investigation 9.4 is turned like this the resulting completed array of numbers appears as a part of a special triangular array of numbers called **Pascal's Triangle**. This array was named after the French philosopher mathematician Blaise Pascal (1623–1662), who discovered many interesting patterns in the array.

The following exercises will provide an opportunity for the reader to search for patterns in Pascal's Triangle.

EXERCISE SET 9.4

1. One arrangement for Pascal's Triangle is

$$
\begin{array}{c}
1 \\
1 \quad 1 \\
1 \quad 2 \quad 1 \\
1 \quad 3 \quad 3 \quad 1 \\
1 \quad 4 \quad 6 \quad 4 \quad 1 \\
\cdot \\
\cdot \\
\cdot
\end{array}
$$

Observe the relationship between each entry and the two entries directly above it. Use this relationship and complete an additional five rows of the array.

2. What interesting number sequences can you find in the diagonals of Pascal's Triangle?

3. An interesting pattern emerges when the numbers in each row of Pascal's Triangle are added.

   a) Complete four more rows of this table.

   b) Without adding the numbers, can you determine the sum of the numbers in the tenth row?

   c) Write a formula for determining the sum of the numbers in the nth row of Pascal's Triangle.

Row number	Sum of the numbers in this row
1	1
2	2
3	
4	
5	
6	
.	
.	
.	
n	

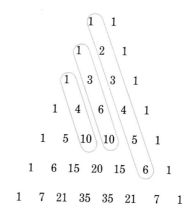

4. It is also easy to find the sum of numbers along a diagonal of Pascal's Triangle.

   a) Find the sum of the numbers in each gray loop.

   b) Do you find these sums in the triangle?

   c) Form a generalization.

5. From the array in Exercise 1, observe that the 1st, 2nd, and 4th rows of Pascal's Triangle contain only odd numbers.

   a) What is the next row that contains only odd numbers?

   b) Can you find a pattern for determining which rows contain only odd numbers?

6. Draw twelve rows of Pascal's Triangle for each part of this exercise.

   a) Circle all even numbers in an array. What patterns do you see?

   b) Circle all multiples of 5 in an array. Do you see any patterns?

   c) Circle all multiples of 3. What patterns emerge?

7. As the second number in the row, the number 5 is a divisor of each other number in its row (except the number 1).

   a) What other second numbers in rows have this property?

   b) Describe a way to determine, simply from observing the second number, whether or not the number divides each other number in the row (except the number 1) evenly.

8. The sums of the numbers in these first seven special diagonals of Pascal's Triangle are given.

   a) Find a pattern in this sequence of numbers and guess the next three numbers in the sequence. [*Hint*: Consider the two numbers preceding any number in the sequence.]

   b) Draw more rows of Pascal's Triangle and check your guesses in part (a).

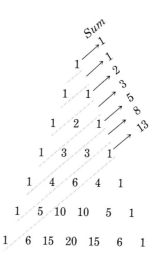

n	Sum of the first n Fibonacci numbers
1	1
2	2
3	4
4	7
5	
6	
7	

9. The sequence of numbers you discovered in Exercise 8 is called the **Fibonacci Sequence**, named after the great medieval mathematician Leonardo of Pisa (called Fibonacci).

a) Write the first 15 Fibonacci numbers.

b) Study these equations for finding the sums of Fibonacci numbers. Complete the next three lines of the table.

$$1 = 2 - 1 = 1$$
$$1 + 1 = 3 - 1 = 2$$
$$1 + 1 + 2 = 5 - 1 = 4$$
$$1 + 1 + 2 + 3 = 8 - 1 = 7$$
$$1 + 1 + 2 + 3 + 5 = 13 - 1 = 12$$

*10. Given banks of light switches with 2, 3, 4, 5, and 6 switches each, an interesting question is, how many different ways are 2 switches on? 3 switches on?

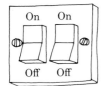 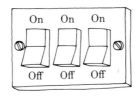 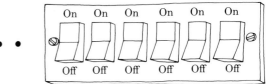

Record in the appropriate entry in the table the number of different ways of having the required number of "on" switches.

Number of switches in bank \ Number of On switches	1	2	3	4	5
1	1	0	0	0	0
2	2	1			
3	3				
4					
5					
6					

*11. Suppose you have a set of five elements, a, b, c, d, e.

    a) How many subsets containing 0 elements does this set have?

    b) How many different subsets containing one element does the set have?

    c) How many different two element subsets does the set have? 3 element subsets? 4 element subsets? 5 element subsets?

    d) Do the numbers you found in parts (a), (b), and (c) have any relationship to Pascal's Triangle?

    e) Form a generalization about situations like these.

BIBLIOGRAPHIC REFERENCES

[6, 26–33, 50–61, 204–215], [17, 52–59], [20, 12–20], [34, 125–127, 131–146], [45], [46], [82], [89]

PEDAGOGICAL ACTIVITIES FOR THE TEACHER

1. Create an investigation card which is appropriate for a level of your choice in which the student is led to a discovery of some pattern related to the triangular and square numbers.

2. Read a book such as *String Sculpture* by John Winter (Creative Publications) and (a) write directions for the process of curve stitching, (b) make some special patterns which could be used by students to make interesting curve stitching designs, and (c) present techniques you use to motivate students to become involved in a curve-stitching project.

3. A notebook or card file of ideas for teaching mathematics is a valuable aid to teachers at every level. Not only should teachers continue to collect such ideas, but persons preparing to be teachers should start this collection as soon as possible. To this end consult the book by Don Cohen, *Inquiry in Mathematics via the Geoboard* (Walker Educational Book Corporation), and select at least three patterns that students at a given level can discover using a rectangular or circular geoboard.

4. Complete the activities presented in the Madison Project "Shoeboxes" (available from Math Media, Inc.) and list the generalizations you have discovered. Discuss the level at which students might become involved with these activities.

10
Measuring with Metric Units

A wall chart on the history of measurement is available from Ford Motor Company. Students enjoy this chart because it colorfully depicts the use of the human body in selecting standard units of measure.

The upper part of this painting from an Egyptian tomb shows the "rope stretchers" engaged in surveying. Other aspects of measurement appear in the lower portion.

INTRODUCTION

From the time of the baked clay tablets of Mesopotamia (dated prior to 3000 B.C.) to the present day, we find records of the importance of measurement. The Babylonians and Egyptians and their contemporaries appeared to view all of geometry as related to the practical process of measurement. In fact, as mentioned in Chapter 1, the word "geometry" (geo-metry, meaning "measurement of the earth") probably originated in these cultures.

Lord Kelvin, whose name is used for the basic scientific unit of temperature in the metric system, asserted, "When you can measure what you are speaking about and express it in numbers, you know something about it; but when you cannot measure it, when you cannot express it in numbers, your knowledge is of a meagre and unsatisfactory kind" (from *Popular Lectures and Addresses,* Macmillan, 1891).

Whether one agrees with Kelvin's observation or not, it does highlight an important practical physical process. This process of measurement, synonymous with "geometry" in early cultures and still of tremendous practical importance in today's world seems an appropriate place for us to begin.

Thebes: Tomb 38 – Field Measurement

537 Copyright by Lehnert & Landrock, Cairo

THE MEASUREMENT PROCESS

The display in Fig. 10.1 provides a concise description of the process of measurement. Investigation 10.1 will provide an opportunity for exploration of some of the key elements of this process.

The measurement process

1. Choose an object or event and select a measurable property p (such as length, mass, time, capacity, etc.) to be measured.	2. Select an appropriate unit, depending upon the precision of measurement required. 3. Use as many units as necessary, in an orderly way, to "reach the end", "cover", "fill up", or otherwise assess the property p. 4. Ascertain the number of units used.	5. Assert the number of units used to be the measure of p.

Figure 10.1

Before starting the investigation, you might spark your creativity by reading these excerpts from a story by Alexander Calandra in the December 21, 1968 issue of the *Saturday Review*. In response to the problem, "Show how it is possible to determine the height of a tall building with the aid of a barometer," a student made the following responses:

> "Take the barometer to the top of the building, attach a long rope to it, lower the barometer to the street, and then bring it up, measuring the length of the rope. The length of the rope is the height of the building." . . .

> "Take the barometer to the top of the building and lean over the edge of the roof. Drop the barometer, timing its fall with a stopwatch. Then, using the formula $S = \frac{1}{2}at^2$, calculate the height of the building." . . .

> . . . "you could take the barometer out on a sunny day and measure the height of the barometer, the length of its shadow, and the length of the shadow of the building, and by the use of a simple proportion, determine the height of the building." . . .

> . . . "There is a very basic measurement method that you will like. In this method, you take the barometer and begin to walk up the stairs. As you climb the stairs, you mark off the length of the barometer along the wall. You then count the number of marks, and this will give you the height of the building in barometer units. A very direct method.

"Of course, if you want a more sophisticated method, you can tie the barometer to the end of a string, swing it as a pendulum, and determine the value of 'g' at the street level and at the top of the building. From the difference between the two values of 'g,' the height of the building can, in principle, be calculated.

"Finally," he concluded, "there are many other ways of solving the problem. Probably the best," he said, "is to take the barometer to the basement and knock on the superintendent's door. When the superintendent answers, you speak to him as follows: 'Mr. Superintendent, here I have a fine barometer. If you will tell me the height of this building, I will give you this barometer.' " . . .

Investigation 10.1

Work in groups on at least one of the following questions. Be creative!

A. In how many different ways can you measure a ball?

B. Which measurable properties of the human body can you measure with

   a) a large piece of graph paper?

   b) a piece of string?

   c) two containers, some water?

   d) a stop watch?

   e) a balloon?

As we further consider the description of the measurement process in Fig. 10.1, we notice the following key elements in the process.

1. Choosing measurable properties

2. Selecting appropriate units

3. Developing and using measuring devices

4. Associating a number with the measurable properties

In the investigation, for example, a person may have chosen to measure the distance around the ball (the choice of the measurable property) with a centimeter tape measure (the choice of a unit and a measuring device). As the tape is wrapped around the ball (the use of the measuring device) the tape meets at 28 and we say the circumference of the ball is 28 centimeters (the association of a number with the measurable property).

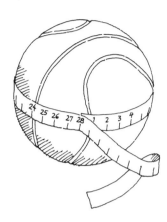

Thus the measurement process is essentially a physical means of comparing the measurable property with a unit of measure. A device is often used to aid in this comparison and a number is used to describe this comparison. The basic idea is further illustrated in Fig. 10.2. The reader should focus on each aspect of the measurement process in these illustrations.

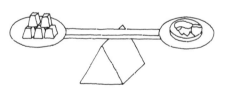

The nickel balances 5 standard gram units. The mass of the nickel is 5 grams.

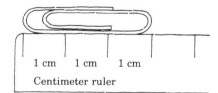

The paper clip is as long as 3 centimeter units. The length of the paper clip is 3 centimeters.

1 cm   1 cm   1 cm

Centimeter ruler

Figure 10.2

The following exercises will provide an opportunity for further exploration of the key elements of the measurement process.

EXERCISE SET 10.1

1. a) List as many different measurement devices as you can without consulting a reference. Describe the measurable property which is measured by each of these devices.

   b) List as many commonly used units of measure as you can without consulting a reference.

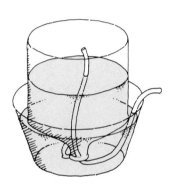

2. A jug is partially filled with water, turned upside down in another container of water, and a rubber tube is inserted as shown. If the person blows on the tube, the water level in the jug will recede.

   Discuss how the equipment pictured could be used as a *device* to measure some *measurable property* of the human body.

   Choose a *unit of measure* and describe how to complete the measuring device so that it will easily help a person *associate a number with the measurable property*.

3. Describe as many measurable properties as you can think of for which a "tire pump" could be used as a measuring device.

4. Consider each of these measurable properties:

   a) speed of a runner,

   b) force of gravity near earth's surface,

   c) length of a swinging pendulum,

   d) distance between two geographical points,

   e) temperature of a hot plate.

   For which of these measurable properties could a "stop watch" be used as the measuring device which associates a number with the measurable property?

5. List some of the decisions which involve a measurement process which one makes when attending a professional athletic contest.

THE NEED FOR STANDARD UNITS IN MEASURING

A basic step in the measurement process is the choice of a unit of measure. The following investigation may help the reader capture the spirit of this importance and help establish criteria for choosing units.

It is useful to encourage students to invent their own unit of measure using an object such as a shoe, a pencil, or a part of their body. If a group of students use the width of the palm of their hand as a unit and find they get different numbers for the measure of the same thing, the need for the standard unit becomes vivid.

Investigation 10.2

"And there went out a champion out of the camp of the Philistines, named Goliath of Gath, whose height was six cubits and a span" (I Samuel 17:4)

A. Using your cubit and span, cut a string that is as long as Goliath was tall. Compare your string length with the strings of others in your group. Discuss any difficulties that might arise from choosing and using units determined by each person's own body.

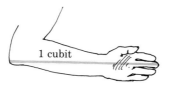
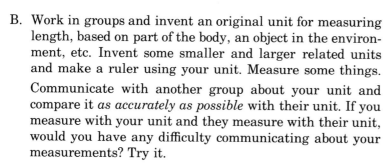
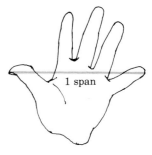

B. Work in groups and invent an original unit for measuring length, based on part of the body, an object in the environment, etc. Invent some smaller and larger related units and make a ruler using your unit. Measure some things.

Communicate with another group about your unit and compare it *as accurately as possible* with their unit. If you measure with your unit and they measure with their unit, would you have any difficulty communicating about your measurements? Try it.

Discuss any difficulties that might arise if each nation had a local unit of measure different from others.

The investigation suggests that we have complete freedom in choosing units of measure, but that indiscriminate use of this freedom can often make accurate communication of the results of our measurements very difficult.

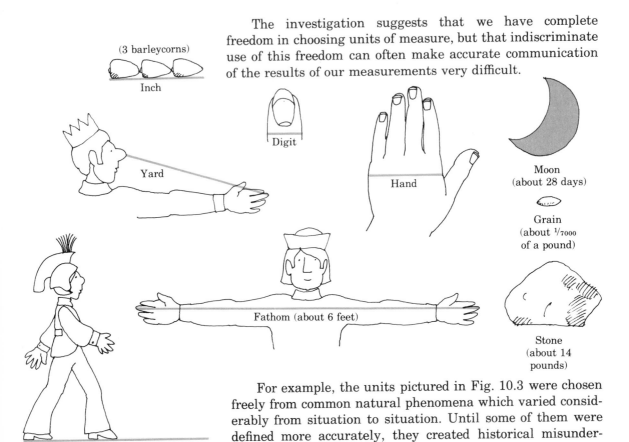

(3 barleycorns)

Inch

Digit

Yard

Hand

Moon
(about 28 days)

Grain
(about $1/7000$
of a pound)

Fathom (about 6 feet)

Stone
(about 14
pounds)

Pace (Roman double step)
(about $1/1000$ of a mile)

Figure 10.3

For example, the units pictured in Fig. 10.3 were chosen freely from common natural phenomena which varied considerably from situation to situation. Until some of them were defined more accurately, they created historical misunderstandings among both people and nations which had far-reaching implications.

As commerce developed and trade between nations became possible over longer distances, sociological and economic issues motivated a search for highly standardized, uniformly accepted units of measure. Scientists needed to communicate ideas without undo confusion, and were among the first groups to adopt a common system of measurement units. Manufacturers and other tradesmen followed in many countries, in order to eliminate misunderstandings and to provide for easily interchangeable parts on the products traded. A popular system which has gained almost worldwide acceptance will be described in the next section.

THE METRIC SYSTEM OF MEASUREMENT UNITS

In a recent year it was recorded that almost all the nations in the world use a standard system of measurement called the **International Metric System** or *The Système International d'Unités* (SI System).

The metric system has been accepted as the standard system of units because it is basically simple. As Fig. 10.4 indicates, the *meter, liter,* and *gram*, together with a few related units, are sufficient for everyday practical affairs.

This system is simple because of the use of a limited number of basic units, the use of our familiar decimal system of numeration, and by repeated use of a few prefixes and basic units to name smaller or larger related units.

For example, distances on the earth are usually measured using a unit 1000 meters long. This unit is named using the prefix "kilo" and the basic unit "meter." Thus the distance from New York to San Francisco is about 4840 *kilometers*.

The width of photographic film is measured using a unit 1/1000 of a meter long. This unit is named using the prefix "milli" and the base unit "meter." Thus the width of a popular size of film is 16 *millimeters*.

The most commonly used metric system prefixes are given in Table 10.1. For a more complete list, see Table 10.3.

From the basic units meter, liter, and gram and these prefixes, other useful units are named. Those units most commonly used are described briefly below and will be developed more specifically in later sections of this chapter.

Students develop attitudes about mathematics and measurement which often correlate highly with the attitude of the teacher toward the subject. Books such as *A Metric America – A Decision Whose Time Has Come* (National Bureau of Standards, 1971) and the April 1973 issue of *The Arithmetic Teacher* not only provide information about the metric system but some cogent reasons why this system should gain acceptance.

← 16 mm →

Table 10.1 Most Commonly Used Metric System Prefixes

Prefix	Meaning
kilo-	1000 (one thousand)
centi-	0.01 (one hundredth)
milli-	0.001 (one thousandth)

Length		Capacity	
Meter	about 1.1 yard	Liter	about 1.1 quart
Kilometer	1000 meters, about 0.6 mile	Milliliter	0.001 liter, five of them fill a teaspoon
Centimeter	0.01 meter, about the width of a paperclip	**Weight**	
Millimeter	0.001 meter, about the thickness of a paperclip wire	Gram	about the same weight as 2 raisins or a large paperclip
		Kilogram	1000 grams, about 2.2 pounds

All You Will Need to Know About Metric

(For Your Everyday Life)

10

Metric is based on Decimal system

The metric system is simple to learn. For use in your everyday life you will need to know only ten units. You will also need to get used to a few new temperatures. Of course, there are other units which most persons will not need to learn. There are even some metric units with which you are already familiar: those for time and electricity are the same as you use now.

BASIC UNITS

METER: a little longer than a yard (about 1.1 yards)
LITER: a little larger than a quart (about 1.06 quarts)
GRAM: about the weight of a paper clip

(comparative sizes are shown)

1 METER

1 YARD

COMMON PREFIXES
(to be used with basic units)

Milli: one-thousandth (0.001)
Centi: one-hundredth (0.01)
Kilo: one-thousand times (1000)

For example:
1000 millimeters = 1 meter
100 centimeters = 1 meter
1000 meters = 1 kilometer

25 DEGREES FAHRENHEIT

1 LITER   1 QUART

OTHER COMMONLY USED UNITS

Millimeter:	0.001 meter	diameter of paper clip wire
Centimeter:	0.01 meter	width of a paper clip (about 0.4 inch)
Kilometer:	1000 meters	somewhat further than ½ mile (about 0.6 mile)
Kilogram:	1000 grams	a little more than 2 pounds (about 2.2 pounds)
Milliliter:	0.001 liter	five of them make a teaspoon

OTHER USEFUL UNITS

Hectare: about 2½ acres
Tonne: about one ton

25 DEGREES CELSIUS

TEMPERATURE
degrees Celsius are used

°C	−40	−20	0	20	37	60	80	100
°F	−40	0	32	80	98.6	160		212

water freezes   body temperature   water boils

1 KILOGRAM   1 POUND

Fig. 10.4 A copy of this information bulletin and further information about the metric system is available from the Metric Information Office, National Bureau of Standards, Washington, D.C. 20234.

Table 10.2 Basic Units in the Metric System

Measurable property	Metric unit	Symbol for the unit
Length	meter	m
Mass	kilogram	kg
Time	second	s
Molecular substance	mole	mol
Temperature	Kelvin	K
Electrical current	ampere	A
Luminosity	candela	cd

Not only is the metric system very simple for everyday use, but it is very powerful for use in scientific endeavors. All measurements, for example, including intricate ones required in scientific explorations, can be made using the seven basic units listed in Table 10.2 or units derived from them. Examples of derived units are those used in measuring capacity, area, volume, velocity, force, work, frequency, electrical resistance, etc. The seven basic units are standardized to a high degree of accuracy based on scientific criterion. For example, the meter is defined to be 1,650,763.73 wavelengths in a vacuum of the orange-red line of the spectrum of the element Krypton 86. This length never varies and can be replicated in laboratories all over the world.

These units, together with the prefixes listed in Table 10.3, are sufficient for any desired measurement.

As an example of the use of very small units, physicists have found ways to measure minute particles less than one trillionth of a meter in diameter for use in particle accelerators. One unit used to record measurements such as this is named using the prefix "pico" and the base unit "meter." Thus the particle is measured in *picometers*.

At the other end of the continuum of measurement, astronomers measure distances to galaxies billions of light years away. (A light year is approximately 9.6 quadrillion meters.) Thus the prefix "tera" and the base unit "meter" might be combined to name a unit, the *terameter*, which could be used to record some of the measurements in astronomy.

Table 10.3 List of Prefixes
Used in the Metric System

Prefix	Meaning

This section of the table contains the prefixes which are used to name units which might be used in astronomy to measure *long distances.*

tera	one trillion times
giga	one billion times
mega	one million times

This section of the table contains the prefixes that are used to name units that can be used in *practical measurement.* Some of these prefixes aren't used very often. Others, such as *kilo-, centi-,* and *milli-,* are frequently used to name common metric system units.

kilo	one thousand times
hecto	one hundred times
deka	ten times
deci	one tenth of
centi	one hundredth of
milli	one thousandth of

This section of the table contains the prefixes that are used to name units that might be used in science to measure *very small lengths.*

micro	one millionth of
nano	one billionth of
pico	one trillionth of
femto	one quadrillionth of
atto	one quintillionth of

Clearly, the present-day measurements, the communications about these measurements, and the use of products built using these measurements by peoples in different countries require a carefully standardized system of units with subunits sufficient to attain the precision of measurement desired. This requirement is met in the metric system, which is the topic of the exercises which follow and which will be discussed further in later sections.

EXERCISE SET 10.2

1. a) Without referring to any table, name as many of the metric system prefixes as you can and give their meaning. If you do not already know the prefixes in the shaded part of Table 10.3, memorize them now. (You may want to make cards with the prefix on the front and the meaning on the back to help you in the memorization process.)

b) If you feel a teacher should know them, memorize the other prefixes in Table 10.3.

2. Even though the metric prefixes are not used with the *day* as unit to indicate time, complete the following as if they were, in order to increase your familiarity with the prefixes.

a) How long is a *deci*day? How long is a *centi*day?

b) Which is longer—a *milli*day or a minute? How much longer?

c) How long is a *deka*day?

d) Which is longer—a *kilo*day or a year? How much longer?

3. How much is a *milli*buck? a *centi*buck? a *deci*buck? a *deka*buck? a *kilo*buck? a *milli*grand? (a "grand" is $1000)

4. Study the example of the basic unit in the metric system. Then estimate the measurement described.

a) Ten straight pins weigh about 1 *gram*. Estimate the weight of a pencil in grams.

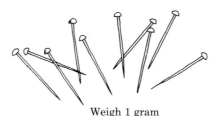

Weigh 1 gram

b) A *meter* is about as long as a long pace. Estimate the length of a city block in meters.

c) A *second* is about the length of time it takes to say "one thousand one." Estimate the time it takes to write your name in seconds.

d) A *liter* is about 1.1 quart. Estimate the number of liters it would take to fill a 10 gallon gasoline tank.

e) °C (degree Celsius) is a commonly used metric unit of temperature which is derived from the basic Kelvin unit of temperature. Water freezes at 0° Celsius and boils at 100° Celsius. Estimate the °Celsius of normal body temperature.

Meter

Most children before the age of six or seven are not ready to measure because they are still operating at the topological level and are not aware that a stick or ruler remains the same length regardless of a change in position. For example, if two sticks the same length are moved as shown in Fig. 1 below, the child will assert that Stick *B* is the longer. If they are moved as in Fig. 2, they will assert that Stick *A* is the longer.

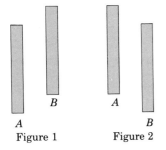

Figure 1

Figure 2

This happens because the child looks only at the endpoint which is now further away. They do not consider the rigidity of the stick and the movement of the other endpoint. Piaget has observed that the necessary concepts for length measurement are achieved on the average around eight years old.

MEASURING LENGTH

The physical process of measurement described at the beginning of this chapter suggested that we first choose a "measurable property" of an object. When this property can be loosely described as "extension in one direction," we are meaning **length**. To measure length, we choose a unit which has this property and compare it with the object to be measured. This is usually done by placing copies of the unit end to end along the object until we reach or surpass the end, and counting the number of units used. The unit of length is arbitrarily chosen. To help children understand this idea, we encourage them to use unorthodox units, such as those shown here, in their initial experiences with the measurement process.

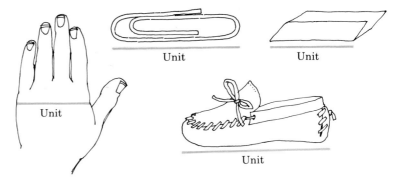

From these experiences children recognize the need for a standard, accepted unit that is the same in all situations.

The basic standard unit of length used in the metric system is the **meter**. Tables 10.4 and 10.5 show this unit and the subunits generated by using some of the standard metric prefixes.

Table 10.4 Some Commonly Used Metric Units of Length

meter	The room is 5 meters wide.
*kilo*meter	We traveled 650 kilometers in one day.
*centi*meter	My waist measures 75 centimeters.

Table 10.5 A More Complete Table of Metric Units for Measuring Length

Unit	Relation to basic unit
*kilo*meter (km)	1000 meters
*hecto*meter (hm)	100 meters
*deka*meter (dam)	10 meters
meter (m)	Basic unit
*deci*meter (dm)	.1 meter
*centi*meter (cm)	.01 meter
*milli*meter (mm)	.001 meter

The pictures below compare the meter with common objects.

This distance on a person might be close to 1 meter.

A baseball bat might be just a little shorter than a meter.

A guard on a university basket ball team might be 2 meters tall.

A **decimeter** is 1/10 of a meter. An orange Cuisenaire rod, used in many classrooms, is 1 decimeter long.

1 decimeter

A **centimeter** is 1/100 of a meter long. These pictures compare this unit with common objects.

The width of a piece of chalk is about 1 centimeter.

The Cuisenaire unit cube is 1 centimeter on each edge.

The width of the base of the pointer finger is about 1 centimeter.

The diameter of the head of a thumbtack is about 1 centimeter.

A **millimeter** is 1/1000 of a meter. The edge of the cardboard back from a tablet is about 1 millimeter thick. So is the wire of a paper clip or the thickness of a dime.

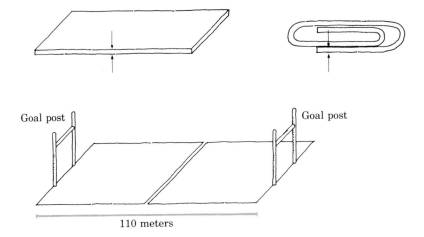

Goal post   Goal post

110 meters

From the goal post at one end of a football field to the goal post at the other end is approximately 110 meters.

A **kilometer** is a unit 1000 meters long. Thus nine football fields (including endzones) laid end to end would extend about 1 kilometer. A train of about 72 cars would be close to 1 kilometer long.

The following investigation will help you develop a better feeling for the size of each of these units.

Investigation 10.3

Use the descriptions of the metric units of length given above and *no calculations or comparisons with other standard units* to individually estimate the following. Then compare your estimates with others. Finally, check your estimates by calculating or measuring.

A. Estimate, in *centimeters*:
  1. your waist measurement
  2. your shoe length
  3. your arm span
  4. another length of your choice

B. Estimate, in *millimeters*:
  1. the width of your hand
  2. the width of a pencil
  3. the length of a paper clip
  4. another length of your choice

C. Estimate in *decimeters*:
  1. your height
  2. the height of your classroom
  3. the dimensions of a chalkboard
  4. another length of your choice

D. Estimate, in *meters*:
  1. the height of a flagpole
  2. the length of your classroom
  3. the length of a tennis court
  4. another length of your choice

E. Estimate, in *kilometers*:
  1. the distance to a nearby city
  2. the height of the tallest of the Rocky Mountains
  3. the height of the highest flying airplane
  4. another distance of your choice

It is crucial that students "get a feel" for metric units. One of the best ways to achieve this is to "think metric." If physical objects are selected to illustrate each metric unit and the student is asked to do extensive estimation of other lengths using these units, the student will gradually develop a real concept of the metric unit. Estimation is the key.

Several devices are available to aid in the measurement of length using the metric units. Some of them are pictured in Fig. 10.5.

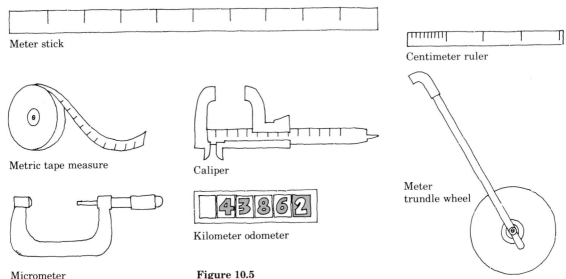

Meter stick

Centimeter ruler

Metric tape measure

Caliper

Kilometer odometer

Meter trundle wheel

Micrometer

Figure 10.5

For children who can count, the trundle wheel is a very nice way to begin to develop the concept of measurement. Instead of having to cope with the concepts necessary to place a ruler end to end a repeated number of times, the child simply pushes the trundle wheel and counts the number of clicks. Thus a number is associated with a distance even though the child may not yet have the concepts necessary to effectively use a ruler to find the number.

The reader should develop an understanding of how each of these devices is used.

Decimals are often used when expressing practical measurements of length in metric units. For example, a person might say, "I am 1.73 meters tall." This statement is easily interpreted as "I am 1 meter and 73 centimeters tall." It might also be interpreted as "I am 1 meter, 7 decimeters, and 3 centimeters tall," but this interpretation is not usually made in practical situations.

Thus the use of the decimal system of numeration in the metric system makes interpretation of measurements quite easy. Since each metric unit of length is 10 times as large as the next smallest unit (check Table 10.5 to see that this is true), the decimal system also makes it convenient to convert from one unit of length to another. This is done by simply moving the decimal point to the left or right. Study this example, reading from left to right and vice versa.

.732 km = 7.32 hm = 73.2 dam = 732. m = 7320. dm
= 73200. cm = 732000. mm

EXERCISE SET 10.3

1. Find at least one common object or distance not mentioned in this text with each of the following lengths:

   a) 1 millimeter   b) 2 millimeters

   c) 5 millimeters   d) 1 centimeter

   e) 2 centimeters   f) 5 centimeters

   g) 1 decimeter   h) 1 meter

   i) 2 meters   j) 5 meters

   k) 1 kilometer   l) 2 kilometers

2. Draw segments which you *estimate* to be each of the following lengths. Then measure the segments and find out by how many millimeters your estimate was off. The total number of millimeters "off" is your "score." Compare your score with others.

   a) 7 cm   b) 3.5 cm   c) 2 cm   d) 8 mm

   e) 10 cm   f) 5.2 cm   g) 4.8 cm   h) 9.3 cm

3. Draw a "broken line" which you estimate to have a total length of 1 meter on a sheet of notebook paper. Measure the line to the nearest centimeter. By how many centimeters did the length of the line differ from 1 meter?

4. Choose an appropriate metric unit and estimate the following. Check your estimate by measuring if possible.

   a) the diameter of a penny

   b) the thickness of a nickel

   c) the length of a new lead pencil

   d) the distance around your wrist

   e) the length of a city block

   f) the distance the winner travels in the Indianapolis 500

   g) the length of your "cubit"

   h) the width of a wristwatch

   i) the length of a study desk

5. Choose a pace which you think is 1 meter long and step off an estimated 10 meters. Measure the actual distance you stepped off and calculate the average length of your pace. By how much did your pace differ from 1 meter?

6. Estimate these speed limits in kilometers per hour.

   a) in a school zone

   b) in a residential area

   c) on a super highway

   d) down main street

   e) on a superhighway for cars with trailers

7. A trip of 5 miles equals a trip of 8 kilometers. Use only this information to give a close estimate of the length of these trips in kilometers.

   a) Chicago to Denver (1038 mi)

   b) Chicago to Washington (698 mi)

   c) Los Angeles to Houston (1540 mi)

   d) Houston to St. Louis (821 mi)

   e) Detroit to Boston (735 mi)

   f) Washington to New York (233 mi)

8. Complete each statement.

   a) 1 m = __ dm   b) 1 m = __ cm

   c) 1 m = __ mm   d) 1 dm = __ m

   e) 1 cm = __ m   f) 1 mm = __ m

   g) 1 cm= __ mm   h) 1 dm = __ cm

   i) 1 mm = __ cm   j) 1 cm = __ dm

   k) 1 m = __ km   l) 1 m = __ hm

9. Complete the following.

   a) 867 cm = __ m   b) 26,428 m = __ km

   c) 632 m = __ cm   d) 367 mm = __ m

   e) 867 m = __ hm   f) 647 cm = __ dm

10. Describe a simple rule for converting from any unit in Table 10.5 to any other unit.

11. Devise a way to use each of the following devices to measure the approximate length of this open curve. Try each way and compare your answers.

a) a piece of string and a metric ruler

b) a compass (or a pair of dividers) and a metric ruler

c) a nickel and a metric ruler

12. a) Measure the length of each hypotenuse in millimeters.

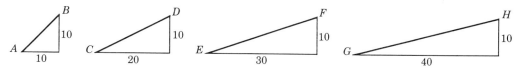

b) Find the actual length using the Pythagorean Theorem. How do your lengths compare?

13. Follow these directions to construct a Golden Rectangle. (See Appendix A for a review of basic ruler and compass constructions.)

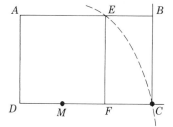

1. Construct a square *AEFD*.

2. Construct the midpoint *M* of *DF*.

3. Extend *DF*.

4. With center *M* and radius *ME*, draw an arc crossing *MF* extended at *C*.

5. Construct a perpendicular to *DC* at *C*.

6. Extend *AE* to intersect the perpendicular at *B*.

7. *ABCD* is a Golden Rectangle.

   a) Measure the length and width of your rectangle to the nearest millimeter.

   b) Find the ratio of the length of the width. By how much does this ratio differ from the Golden Ratio (approximately 1.618)?

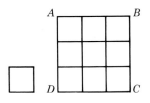

Unit   The area of *ABCD* is 9

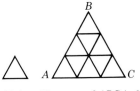

Unit   The area of *ABC* is 9

MEASURING AREA

When the measurable property can be loosely described as "the extent of surface enclosed," we are measuring the **area** of the surface.

Children often think of area in terms of the "amount covered" and count the number of unit squares needed to cover a surface to estimate its area. As with length, the unit of area is arbitrary, and other figures, such as triangles, are often used as units to illustrate this point.

The following investigation will provide some experences which give further meaning to the idea of measuring area.

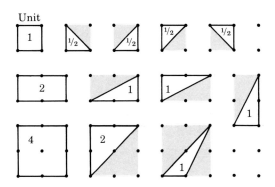

Investigation 10.4

If the unit of area is as shown, the areas of the other regions are as indicated. Be sure you see why. (The shading might help.)

Use these ideas to complete the following.

A. On a 5 × 5 geoboard or a 5 × 5 dot paper grid, seven **squares**, each with different whole number areas, can be shown. How many of these can you find and show on dot paper? (The vertices of the squares must be points on the grid.)

B. Can you find triangles with areas ½, 1, 1½, 2, 2½, 3, 3½, 4, 4½, 5, 5½, 6, 6½, 7, 7½, 8 on 5 × 5 dot paper? Show each one you find.

C. Make a figure on 5 × 5 dot paper for which it is difficult to find its area. Challenge another member of your group to determine the area of your figure.

Thus, at the beginning stages, the process of measuring area is accomplished by counting the number of units, half units, etc. that are needed to "cover up" a region. Children sometimes draw a grid on a sheet of plastic and place a frame around it to make a device to help them find the area of a fig-

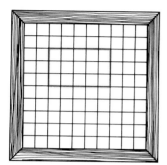

ure. This device is placed over a region, such as the gray rectangle, and the units needed to cover the region are easily counted.

In the metric system the basic unit of area is the **square meter**. For classroom work, the square centimeter or the 2-centimeter square are useful units of area.

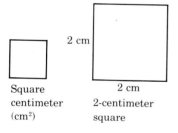

Square centimeter (cm²)

2 cm
2 cm
2-centimeter square

Small land areas are usually measured using a square unit 10 meters on each side (a square dekameter). This unit is called an **are** (pronounced "air") for simplicity. Larger land areas are measured using a square unit 100 meters on each side (a square hectometer).

A Piagetian test a teacher might use to see whether a young child has developed the conservation ideas necessary for beginning to measure area is as follows: Two rectangular sheets of cardboard colored green, both the same shape and size, are presented. A model of a cow is placed on each and the child is told that the farmers who own the pastures have decided to put a house on each. A model of a house is placed near the middle of one pasture and in the corner of the other, and the child is asked whether the two cows still have the same amount of grass to eat. Houses are added in pairs, one to each pasture, but with the houses on one pasture closer together than the houses on the other. The question is repeated. If the child knows that when the pastures contain the same number of houses, regardless of their arrangement, the grass areas must be the same, the child is becoming aware of the concepts necessary for finding area.

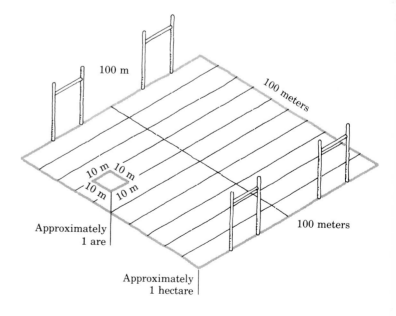

100 m

100 meters

10 m   10 m
10 m   10 m

Approximately 1 are

100 meters

Approximately 1 hectare

This unit is called a **hectare** (pronounced "hectair"). The tables below describe the metric units for area.

Table 10.6 Some Commonly Used Metric Units for Area

are:	My garden plot is 3 ares.
hectare:	A section of land (1 mile square) contains about 260 hectares.

Table 10.7 A More Complete Table of Metric Units for Measuring Area

Unit	Relation to basic unit
square kilometer (km²)	1,000,000 square meters
square hectometer (hm²) or hectare (ha)	10,000 square meters
square dekameter (dam²) or are (a)	100 square meters
square meter (m²)	Basic unit
square decimeter (dm²)	.01 of a square meter
square centimeter (cm²)	.0001 of a square meter
square millimeter (mm²)	.000001 of a square meter

In most practical situations where the area of a region must be found, certain lengths are determined and mathematical formulas are used to calculate the area of the region. Some of these formulas will be developed and used in Chapter 11.

If in a practical situation a person asserts, "The land to be used for this subdivision contains 16.18 hectares," it could be interpreted to mean that the lot contained 16 hectares and 18 ares. This is true because 100 ares make one hectare, just as each metric unit of area is 100 times as large as the next smaller unit (check Table 10.7 to see that this is true). Thus to convert from one unit of area to the next larger unit or next smaller unit, one must move the decimal point 2 places to the left or right.

The following exercises will help clarify the concept of area and the relationships between commonly used metric units.

EXERCISE SET 10.4

1. Measure a square with an area of 1 are (or 1 side of the square if space is limited).

2. A certain city block is 100 meters on each side.
   a) What is the area of this block in hectares?
   b) What is the area of this block in ares?

3. Estimate the area of the following in ares.
   a) a football field   b) a basketball court
   c) the total floor space in a house   d) a small bedroom

4. A lot 35 meters by 35 meters was purchased to serve as a site for a new house. What is the area of this lot in ares?

5. A 40-acre square field is about 400 meters on each side. What is the area of this field in hectares?

6. An area of 10 acres is approximately the same as an area of 4 hectares.
   a) Use this information to check your work in Exercise 5.
   b) Find the area in hectares of a 360-acre farm; a 120-acre subdivision; a 200-acre amusement park.

7. Estimate the area of each figure in square centimeters. Then make a grid on plastic or tracing paper like the one shown here, using the square centimeter as the unit. Use the grid to check your estimate.

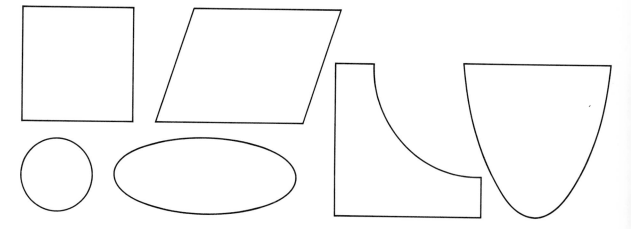

8. Complete the following:

a) There are __?__ square centimeters in a square decimeter.

b) There are __?__ square decimeters in a square meter.

c) There are __?__ square meters in a square deka-meter.

d) There are __?__ square meters in a hectare.

e) There are __?__ square meters in an are.

f) There are __?__ ares in a hectare.

9. Square A was dissected and the pieces reassembled to form rectangle B.

a) What is the area of square A?

b) What is the area of rectangle B?

c) How can you explain the difference in the areas of Fig. A and B?

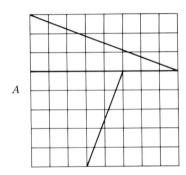

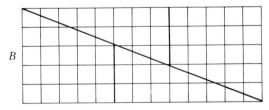

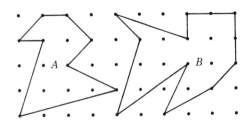

10. Find the area of Figs. A and B shown at the left.

11. a) Draw some figures on a large sheet of paper. Use tracing paper or a grid and devise a way to find the area of these figures using each of the units shown here.

   b) Discuss reasons why the square is usually used as the unit of area.

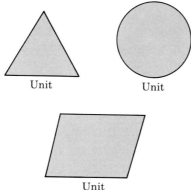

Unit   Unit

Unit

12. The seven pieces of the Tangram puzzle are shown here.

   a) If the area of the small square is 1, what is the area of the large square?

   b) If the area of the large square is 1, what is the area of each of the 7 pieces?

   c) Make up some other area problems like this.

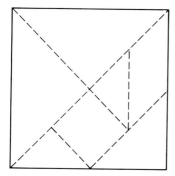

Discuss the dissection puzzles given in Chapter 1 in terms of their providing useful ways to find the areas of certain figures.

MEASURING VOLUME AND CAPACITY

The concepts needed to have an understanding of volume come later than those of length and area. During the middle grades children begin to understand that a stack of blocks 3 × 3 × 4 makes a "house" with the same amount of room as a stack of blocks 3 × 1 × 12. It is only around age eleven or twelve that the child also realizes that the volume of water displaced by these two arrangements is also the same. Earlier the child will assert that the 3 × 1 × 12 arrangement will displace less water because "the house is stretched out lengthwise."

When the "measurable property," as referred to in the description of the measurement process given at the beginning of this chapter, can be described as the "amount of space filled," we are measuring **volume.** Children often think of volume as the "amount the object will hold" or the number of units it takes to "fill the object." Thus the volume of this box, to a child, would be the number of blocks it takes to fill the box.

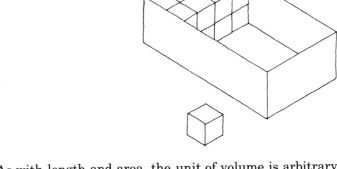

A cubic decimeter contains 1000 cubic centimeters.

As with length and area, the unit of volume is arbitrary. In the metric system the basic unit of volume is the **cubic meter.** The table shows this unit and the units related to it using some of the metric prefixes.

The units in Table 10.8 are not often used in practical measurement situations, but they are important to know about because of their use in science.

Table 10.8 Metric Units for Volume

Unit	Relation to basic unit
cubic kilometer (km³)	1,000,000,000 cubic meters
cubic hectometer (hm³)	1,000,000 cubic meters
cubic dekameter (dam³)	1,000 cubic meters
cubic meter (m³)	Basic unit
cubic decimeter (dm³)	.001 cubic meters
cubic centimeter (cm³)	.000001 cubic meters
cubic millimeter (mm³)	.000000001 cubic meters

The practical counterpart to the more technical measurement of volume is the everyday measurement of "how much some container will hold," called the **capacity** of the container.

Since the cubic meter is somewhat large for ordinary volume measurement and the cubic centimeter is a bit too small, a unit cube 1 decimeter on a side (the cubic decimeter) is often used.

Students need to handle and use containers involving metric units for capacity. Telling and showing is not enough—the child must do in order to really understand.

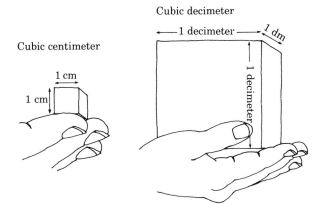

Cubic centimeter

Cubic decimeter

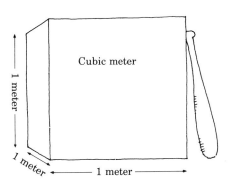

The following investigation will help you focus on a unit that is commonly used to measure capacity.

Investigation 10.5

Devise a way to construct a cubic decimeter container with both inside and outside covered with centimeter graph paper.

Can you answer these questions about your container?

A. How many cubic centimeter containers ( ) full of liquid would it contain?

B. How many of your containers full of liquid would it take to fill a cubic meter tank?

C. How does the capacity of your container compare with the capacity of a quart jar?

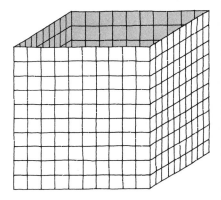

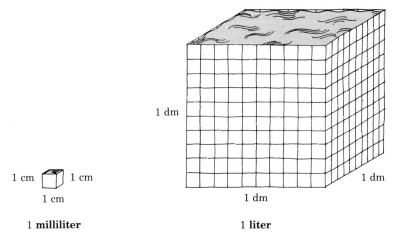

1 dm

1 cm   1 cm

1 cm

1 milliliter   **1 liter**

Since common measures of capacity are often needed, the amount of water an "open top" cubic decimeter will hold is called 1 **liter** of water.

Since it takes 1000 cubic centimeter containers full of water to fill a liter, the capacity of such a container is said to be 1 **milliliter**.

These two units of capacity are the ones commonly used in everyday measurements. The tables below show the related metric units for measuring capacity.

Some Commonly Used Units for Measuring Capacity
liter: This carton of milk holds 2 liters.
milliliter: This can contains 360 milliliters of pop.

Table 10.9 A More Complete Table of Metric Units for Capacity

Unit	Relation to basic unit
kiloliter (kl)	1000 liters
hectoliter (hl)	100 liters
dekaliter (dal)	10 liters
liter (l)	Basic unit = 1 cubic decimeter
deciliter (dl)	.1 liter
centiliter (cl)	.01 liter
milliliter (ml)	.001 liter

As with the measurement of area, the basic means of measuring volume in the preliminary stages is that of counting units. The use of centimeter cubes or centimeter graph paper can sometimes help make this counting process easier, but no standard devices are available to facilitate the counting of the number of units required to "fill an object" to find its volume. Since 1 cubic centimeter equals 1 milliliter, the relationship between volume and capacity is clear.

In practical situations, where the volume of irregular objects is to be measured, the technique of displacement of liquid is sometimes used. In most other cases, certain lengths are measured with standard devices and used in mathematical formulas to calculate the volume. These formulas will be developed and used in Chapter 11.

To convert from one metric unit of volume to another, it should be noted that while units of length differ by multiples of 10, and units of area differ by multiples of 100, the units of volume differ by multiples of 1000. Thus to convert from cubic meters to cubic dekameters, one would multiply by .001, or move the decimal point 3 places to the left. To convert from cubic meters to cubic decimeters one must multiply by 1000, or move the decimal point 3 places to the right.

It should also be noted that the units of capacity, like the units of length, differ by multiples of 10. That is, 10 milliliters = 1 centiliter, 10 centiliters = 1 deciliter, etc.

EXERCISE SET 10.5

1. Name objects with volume of (a) 1 cubic meter, (b) 1 cubic decimeter, (c) 1 cubic centimeter.

2. Name objects with capacity of (a) 1 liter, (b) 1 milliliter.

3. Estimate the capacity of the following in liters or milliliters.

   a) a cup   b) a can of pop

   c) a gallon can   d) a quart jar

   e) a thimble   f) a pint carton

   g) an economy car gas tank   h) a teaspoon

4. Estimate the volume of the following in cubic meters, cubic decimeters, or cubic centimeters.

    a) a classroom   b) a shoebox

    c) a desk drawer   d) a three-drawer filing cabinet

    e) a child's lunch pail   f) a refrigerator

5. Complete the following:

    a) _____ $cm^3 = 1\ m^3$ b) _____ $dm = 1\ m^3$

    c) _____ $mm^3 = 1\ m^3$ d) _____ $m^3 = 1\ km^3$

    e) _____ $dam^3 = 1\ km^3$ f) _____ $hm^3 = 1\ km^3$

    g) _____ $dm^3 = 1\ dam^3$ h) _____ $mm^3 = 1\ km^3$

6. Complete the following:

    a) _____ ml = 1 l   b) _____ cl = 1 l

    c) _____ dl = 1 l   d) _____ l = 1 kl

    e) _____ dal = 1 kl   f) _____ hl = 1 kl

7. Complete the following:

    a) 947 ml = _____ cl   b) 947 ml = _____ dl

    c) 947 ml = _____ l   d) 6.496 kl = _____ hl

    e) 6.496 kl = _____ dal   f) 6.496 kl = _____ l

8. Explore the Soma cube puzzle.

    a) Use measurement of length and decide how to find the volume of the completed cube in cubic centimeters.

    b) Estimate the volumes of the individual pieces.

9. A large cube is made from smaller cubes so that there are three smaller cubes along each edge. This cube is then painted red.

    a) How many of the smaller cubes have red on just 1 face?

    b) How many of the smaller cubes have red on exactly 2 faces?

    c) How many of the smaller cubes have red on exactly 3 faces?

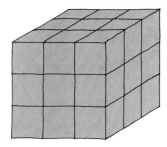

d) How many of the smaller cubes have red on more than 3 faces?

e) How many of the smaller cubes have red on no faces?

*10. What is the largest volume open-top box you can construct using a 16 × 16 square piece of centimeter-ruled graph paper and cutting a square from each corner to make the pattern?

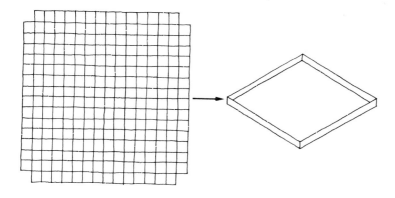

*11. Explain an accurate method for finding the volume of a rock.

MEASURING MASS AND WEIGHT*

When the "measurable property," as referred to in the description of the measurement process given at the beginning of this chapter, can be roughly described by "how much stuff (matter) there is in an object," or, more technically, as "the property of an object which resists acceleration," we are measuring **mass**. When the measurable property can be described as "the amount of attraction between two objects" we are measuring **weight**.

Some interesting experiments for students relating to math as well as to the metric units developed previously in this chapter can be found in the book *The Metric System* (Addison-Wesley, 1974)

---

* These concepts are scientific as opposed to geometrical in nature; consequently, they are not prerequisite to other topics in the text.

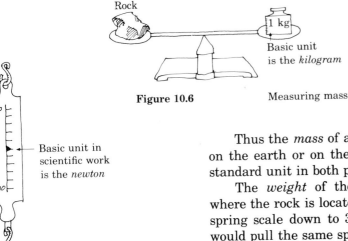

Rock

Figure 10.6

Basic unit
is the *kilogram*

Measuring mass

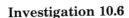

Basic unit in
scientific work
is the *newton*

Rock

Measuring weight

Figure 10.7

To measure mass, we might use a balance and compare the object to be measured with a standard unit (Fig. 10.6). To measure weight, we might use a spring scale and see how much pull is exerted on the object (Fig. 10.7).

Thus the *mass* of a rock remains the same, whether it is on the earth or on the moon, since it will balance with the standard unit in both places.

The *weight* of the rock, however, changes depending where the rock is located. A rock on the earth might pull the spring scale down to 36, while the same rock on the moon would pull the same spring scale down to 6.

This difference between mass and weight is important to understand for scientific purposes, but it doesn't significantly affect practical, everyday measurement. This being the case, we will refer to the "weight of an object" and use the standard units of mass for these measurements.

The following investigation will provide an opportunity for exploration of the basic metric system unit of mass.

Investigation 10.6

For this investigation you will need a balance, the cubic decimeter container you made in Investigation 10.5, and a strong plastic food bag.

Place the plastic bag in the container and fill it with cold water. Place the filled cubic decimeter on the balance.

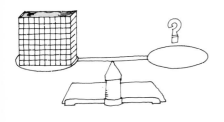

A. Find at least 3 objects or collections of objects that will balance the container of water.

B. About how many objects weighing 1 pound would it take to balance the container of water?

C. Find some metric weights, if possible. Which of these weights or collections of weights will this container of water balance?

The **kilogram**, a basic metric unit of mass equal to the mass of 1 cubic decimeter of water at a temperature of 4° Celsius, is often used in practical measurement. An average man might weigh about 75 kilograms, while a compact car might weigh as much as 900 kilograms.

The **gram** (a unit 1/1000 of a kilogram) is relatively small as indicated by the descriptions and sketches shown at the right.

The following tables describe these and related units.

A large paper clip weighs approximately 1 gram.

Some Commonly Used Metric Units
for Mass (or Weight)

gram:	The small box of raisins weighs 34 grams.
kilogram:	I've lost weight! I now weigh only 70 kilograms.

A sugar cube weighs about 2 grams.

Table 10.10. A More Complete Table of Metric Units for Mass

kilogram (kg)	1000 grams
hectogram (hg)	100 grams
dekagram (dag)	10 grams
gram	basic unit
decigram (dg)	.1 gram
centigram (cg)	.01 gram
milligram (mg)	.001 gram

A nickel weighs about 5 grams.

In practical situations, the word "kilo" is often used instead of "kilogram." If a person said, "I bought a ham which weighed 4.35 kilos," it could be interpreted as "I bought a ham which weighed 4 kilograms and 350 grams." Since 1 kilogram weighs 1000 grams, this interpretation gives a reasonably clear idea as to just how much ham was purchased.

Since each metric unit of mass is 10 times the next smaller unit (check the table above to see that this is true), one can convert from one unit of mass to the next larger or next smaller unit by moving the decimal point one place to the left or one place to the right. For example, 453 grams equals 0.453 kilograms and 453,000 milligrams.

EXERCISE SET 10.6

1. Select objects or collections of objects, weighing
   a) 1 kilogram,   b) 10 grams.

2. What does a cubic centimeter of water weigh?

3. Estimate the weight in grams or kilograms of the following. Check your estimates if possible.

   a) a dime   b) a new piece of chalk
   c) a teaspoon   d) a paper hole punch
   e) a can of pop   f) a small chalkboard eraser
   g) your weight   h) a small ceramic coffee cup
   i) your textbook   j) a penny
   k) a plastic centimeter   l) an unsharpened lead
       ruler   pencil
   m) a ball-point pen   n) a newborn baby
   o) a poodle   p) a "pound of cheese"

4. An object that weighs 11 pounds weighs 5 kilograms. Use
   only this information to find or estimate the weight in kilograms of

   a) a 220-pound man   b) a 4400-pound car
   c) a 55-pound child   d) a 16-pound turkey
   e) a 100-pound sack of   f) a 450-pound decorative
      feed   rock.

5. Which is more, 1.87 kilograms or 1,869 grams?

RELATIONSHIPS AMONG MEASUREMENTS

In addition to length, area, volume, capacity, and weight, the
measure of time and temperature plays an important role in
science and in everyday affairs.

The unit of time, the **second**, is customarily used and is
quite familiar. The unit of temperature in the metric system
is °**Kelvin**, which is especially suited for scientific work. A
unit that is derived from the °Kelvin unit is the °**Celsius**,
or °C.

The basic units of length, area, volume, capacity, mass, temperature, and time are related in interesting and useful ways. For practical purposes an understanding of the meter, liter, kilogram, °Celsius, and second, and the ability to use them and a few related units in ordinary situations is sufficient. For teachers, however, a deeper understanding of the relationships between these units is very helpful. The following investigation will help you become more familiar with the relationship between time and length.

Investigation 10.7

Use a string and a weight to make a pendulum that has *length* 1 meter. Devise an accurate experiment to determine the length of *time* it takes for 1 swing of the pendulum.

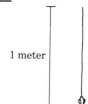

1 meter

Other interesting relationships between the metric units are suggested by the diagram in Fig. 10.8.

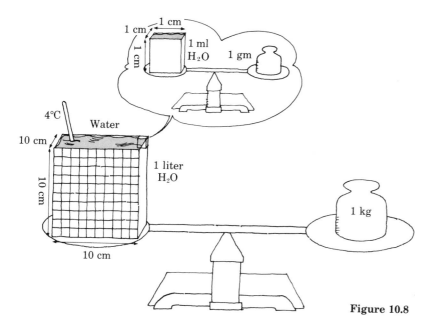

Figure 10.8

The reader should study the diagram and complete the following:

1 cubic centimeter of water at 4°C weighs 1 _____.
1 cubic decimeter of water at 4°C weighs 1 _____.
1 liter of capacity has a volume of 1 _____.
1 cubic centimeter of water has a capacity of 1 _____.
1 milliliter of water at 4°C weighs 1 _____.

We find, then, that the metric system of units is a system in which the basic units are interrelated and in which the interpretation of measurement is simplified through the utilization of the decimal system of numeration.

EXERCISE SET 10.7

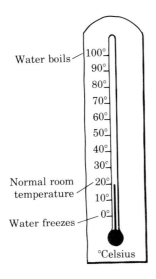

Water boils — 100°
90°
80°
70°
60°
50°
40°
30°
Normal room temperature — 20°
10°
0°
Water freezes —
°Celsius

1. Estimate these temperatures in °Celsius and find their locations on the Celsius thermometer.
    a) normal body temperature
    b) cold pop, without ice
    c) ice cream
    d) hot soup
    e) warm bath water
    f) water from steaming teakettle
    g) hottest earth temperature on record
    h) a cool spring day ("sweater weather")

2. If you know the volume of a water tank in liters, how do you find the weight of the contained water in kilograms?

3. Complete the following:
    a) 1 cubic decimeter of water weighs _____ gram(s).
    b) 1 liter of water weighs _____ gram(s).
    c) 1 liter of water has volume _____ cubic centimeter(s).
    d) 1 liter has volume _____ cubic meter(s).
    e) 1 liter of water weighs _____ kilogram(s).
    f) 1 milliliter of water weighs _____ gram(s).
    g) 1 milliliter of water weighs _____ kilogram(s).
    h) 1 cubic centimeter of water weighs _____ gram(s).

4. The English system of units has been used in the United States for many years, and is of historical interest.

   a) Make a table of length units in the English system and show the comparisons between them. (Use inch, foot, yard, rod, and mile.)

   b) Make a table of area units in the English system and show the comparisons between them. (Use square inch, square foot, square yard, square rod, acre, and square mile.)

   c) Make a table of weight units in the English system and show the comparisons between them. (Use grain, ounce, pound, stone, and ton.)

   d) Make a table of capacity units in the English system and show the comparisons between them. (Use ounce, pint, quart, and gallon.)

5. Compare the relationships in Exercise 4 with the relationships between units in the metric system. What conclusions do you draw?

6. Make a table of conversions from basic metric units to basic English units and vice versa.

7. Work each of these problems. Show your work.

   a) Given a cubical tank 7 feet $8\frac{1}{2}$ inches on a side, find
      i) the volume in cubic feet,
      ii) the volume in gallons,
      iii) the weight of the water in pounds.

   b) Given a cubical tank 3.08 meters on a side, find
      i) the volume in cubic meters,
      ii) the volume in liters,
      iii) the weight of the water in kilograms.

8. Discuss the relative difficulty of computation in the two solutions to the problem in Exercise 7.

9. Work each of these problems. Show your work.

   a) Given a piece of land 500 yards by 500 yards, find the number of acres of land.

   b) Given a piece of land 500 meters by 500 meters, find the number of hectares of land.

10. Discuss the relative difficulty of computation in the two solutions to the problem in Exercise 9.

11. a) If a metric ton (1000 kilograms) costs $150,000, then 1 kilogram costs _____ and 1 gram costs _____.

    b) In the English system, if 1 long ton costs $150,000, then 1 pound costs _____ and 1 ounce costs _____.

    c) Discuss the relative difficulty of computation in the two solutions to the problem in parts a) and b).

12. If you know the miles per gallon which can be driven for a certain car, how can you easily convert this to liters per kilometer?

13. Write appropriate numbers so that this story makes sense.

M(r)(s) Swift had taught _A_ th grade at Chaos School for _B_ years. (S)he was an excellent teacher who enjoyed children and life. (S)he was about _C_ meters tall and weighed _D_ kilograms. (S)he loved to play tennis and golf in the summer, but had difficulty deciding which to do. On a hot _E_ °C day last summer (s)he decided to do both. After 2 sets of tennis and _F_ milliliters of Gatorade, (s)he headed for the golf course. The secret, of course, was to transfer thought from the larger _G_ centimeter diameter, _H_ gram tennis ball to the smaller _I_ millimeter diameter, _J_ gram golf ball. Naturally, (s)he was successful and lofted a respectable _K_ meter drive off of the first tee.

M(r)(s) Swift lived in a large house with _L_ square meters of floor space on a _M_ are lot. (S)he drove a small, economy car, so (s)he didn't need too large a garage. The car *was* economical, getting _N_ kilometers/liter on long _O_ kilometer trips.

M(r)(s) Swift has been dreaming of a large, _P_ hectare wooded area in the country nearby, with a small lake, a cottage, and a cow that gives _Q_ liters of milk a day.

Back to reality, M(r)(s) Swift is looking ahead to teaching school next fall and (s)he is making plans to introduce the metric system to the students. (S)he is a devoted teacher and a "real" person.

BIBLIOGRAPHIC REFERENCES

[30], [49], [76]

PEDAGOGICAL ACTIVITIES FOR THE TEACHER

1. Suppose you as a teacher received a letter from a parent who felt that the U. S. should not change to the metric system because it would cause a lot of trouble, it would cost his business a lot of money, it would be difficult for chefs and real estate agents, and it would be impractical for almost everyone. The parent wonders why you as a teacher are spending so much time on the metric system, why you think it is better, and why parents weren't asked to help make the decision about changing to the new system. Write a letter giving your response to this parent.

2. Here are some reasons which have been given for a change to the metric system. Do you agree or disagree with these reasons? Present your position.

   a) The confusion caused by certain groups in our country using the metric system (certain sports, film manufacturers, pharmaceutical suppliers, the medical profession, the optical manufacturers, etc.) and other groups using the English system would be eliminated. Large corporations and manufacturers are already changing to the metric system and it would be better for the whole country to adopt it with a coordinated plan.

   b) Time would be saved in the elementary school mathematics and science curriculum (some persons estimate as much as two years) because:

   i)   The metric system is easier to learn; fewer, more related units.
   ii) Calculations take less time; mostly whole numbers and decimals instead of fractions and mixed numerals.
   iii) Teaching a dual system of measurement (English for practical use and Metric for science use) would be unnecessary.

iv) Less developmental work with fractions, such as twelvths, thirds, and so on, in complicated operations on mixed numbers would be required.

v) Metric system is more helpful in the cultivation of other important arithmetic skills.

c) Exports are important in maintaining a favorable balance of trade for the United States. The U.S. exportation is handicapped by our production of materials using English measurements for parts to be sold in countries where repairs must be made according to metric specifications. A change to the metric system would allow us to have input into the setting of international standards on industrial products. We put ourselves at a competitive disadvantage in trade by using a measuring system different from that of the world market.

3. Choose an area such as metric length, area, capacity, or mass and write a set of behavioral objectives which describe the outcomes of a unit on the metric system at a given grade level.

4. Develop a learning activity for one aspect of (a) length, (b) area, (c) capacity, (d) mass, using metric system units.

5. Refer to catalogues from producers of commercial materials and prepare a requisition ordering the materials which you believe are needed to do an adequate job teaching the metric system at a selected grade level. After you have prepared the desirable list of necessary materials, assume that your school system can provide only $50 for your classroom. Reduce your list to accommodate this budget restriction.

6. Certain issues are current with regard to the adoption of the metric system. Read some references on the metric system and give your answers to the following questions.

a) Should we teach children to convert from English units to metric units, and vice versa?

b) Should we provide experiences where children compare English units to metric units, and vice versa?

   c) Should we do neither (a) nor (b), but rather immerse
     children in a totally metric curriculum?

   d) How shall parents be informed about the metric
     system?

7. Consult a book such as *Let's Play Games in Metrics* (National Textbook Company, Skokie, Illinois) by George Henderson and Lowell Glunn. After exploring such a book, invent a game of your own which could be used with students at a given grade level. If needed, revise the game so that it will be more appropriate.

8. Devise an interview which could be used with students at a given grade level, or with adults, to determine the extent of their current knowledge about the metric system. Make the interview brief, containing only about four or five questions. Interview various persons and display your collected information graphically.

11
Measurement and Measure

INTRODUCTION

In Chapter 10 we emphasized the measurement process, a process which compares two objects relative to a measurable property and assigns a number. The measurable property is actually a concept. Whether we use a crude measuring device such as a rope or a ruler, or a sophisticated device such as a micrometer, radar, or laser gun, we are comparing two objects relative to some concept. In this chapter we shall study the concepts of the length function, the area function, and the volume function. We shall discuss properties of these functions which can be used to derive formulas with which lengths, areas, and volumes can be computed.

LENGTH

The measurement of length is a process in which we compare two objects and assign a number. The comparison is between the object to be measured and a unit object which in principle can be selected arbitrarily. This process, being physical, is by its very nature an approximation, for a comparison which appears to be "exact" will usually be an approximation when examined under a magnifying glass or a microscope.

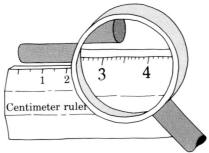

According to some learning theorists the concepts of distance and length are not the same psychologically and must be treated separately. Distance refers to the linear separations of objects; length refers to a property of the objects themselves.

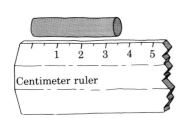

The length of the rod appears to be exactly 3½.

The length of the rod appears to be less than 3½.

To illustrate the process of measuring length consider segment $\overline{AB}$ together with segments labeled a, b, and c. Suppose that we select a as unit length and compare the unit segment a with segment $\overline{AB}$ by laying off segment a onto $\overline{AB}$. We see that two copies of a do not cover $\overline{AB}$ and three copies of a

A _____ B

a _____

b _____

c _____

overlap the end of $\overline{AB}$. We shall describe this observation by borrowing notation from the real number system and writing $2a \leq \overline{AB} \leq 3a$. This estimate of the length of $\overline{AB}$ is seen as a rough estimate and can be improved by using the unit segments b or c. You will do this in Investigation 11.1.

A _____ B

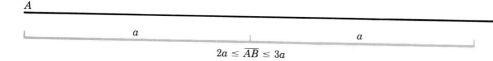

$$2a \leq \overline{AB} \leq 3a$$

Investigation 11.1

A. Using tracing paper, repeat the above estimating process by using unit b and unit c to complete the following:

   __$b \leq \overline{AB} \leq$ __b
   __$c \leq \overline{AB} \leq$ __c

B. Repeat the above process with a centimeter ruler using the common length units of decimeter, centimeter, and millimeter, and complete the following:

   __dm $\leq \overline{AB} \leq$ __dm
   __cm $\leq \overline{AB} \leq$ __cm
   __mm $\leq \overline{AB} \leq$ __mm

C. Compare your results with a partner. Even though this process is an estimating process, do you and your partner agree that the millimeter estimate is exact enough for your purposes?

In part B of this investigation you should have found that:

$$1 \quad \mathrm{dm} \le \overline{AB} \le 2\ \mathrm{dm}$$
$$15 \quad \mathrm{cm} \le \overline{AB} \le 16\ \mathrm{cm}$$
$$153\ \mathrm{mm} \le \overline{AB} \le 154\ \mathrm{mm}$$

Note that we can build on our experience with units gained in Chapter 10 and write these estimates in terms of a single unit of length called a *meter* (m) by using decimal notation. We have:

$$.1 \quad \mathrm{m} \le \overline{AB} \le .2\ \mathrm{m}$$
$$.15 \quad \mathrm{m} \le \overline{AB} \le .16\ \mathrm{m}$$
$$.153\ \mathrm{m} \le \overline{AB} \le .154\ \mathrm{m}$$

At this stage we have estimated the length of $\overline{AB}$ to be .153 m. Whether or not this last estimate is accurate enough depends upon the context in which the measurement takes place. The desired accuracy of a measurement is always a human decision. While you and your partner may have decided that 153 mm was an "exact enough" estimate, an architect or an engineer might require greater accuracy. If that is the case, a smaller unit of length and a more accurate measuring device needs to be chosen and the process must be repeated. For example, if the unit of length called $^1/_{10}$ *millimeter* is chosen, we may find, for segment $\overline{AB}$, that

$$1532\ (^1/_{10}\ \mathrm{mm}) \le \overline{AB} \le 1533\ (^1/_{10}\ \mathrm{mm}),$$

or, equivalently,

$$0.1532\ \mathrm{m} \le \overline{AB} \le 0.1533\ \mathrm{m}.$$

In practice the physical measurement process rather quickly reaches the point which represents the measurement of greatest accuracy for the given measuring device. On the other hand, it is theoretically possible to imagine that this process goes on without end. (See Fig. 11.1.) There are two possible outcomes of the theoretical process.

1. At some stage the point B may fall exactly on a division line of the ruler and we obtain a number called the length meaure of $\overline{AB}$.

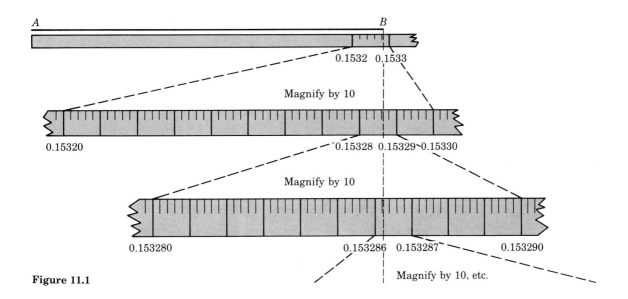

Figure 11.1

2. Point B may not fall on a division point of the ruler no matter how far the process is carried. In this case we obtain an unending decimal numeral which represents a number called the length measure of $\overline{AB}$. (Recall that if this unending decimal numeral is repeating, it represents a rational number. Otherwise it represents an irrational number.)

Summarizing, we see that once we select a unit, the above theoretical process associates to each segment a real number called the length of the segment. Choosing the unit is the determining factor in establishing the rule which associates each segment with a unique number. So the concept of length measure is actually a *function*—a rule which associates a unique number with each segment. To describe the situation pictured here, we write $L(\overline{AB}) = 1.7$ (we read "the length measure of segment AB is 1.7").

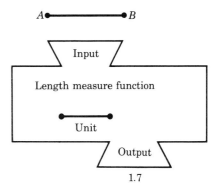

Observe that the availability of irrational numbers for lengths is required in order to assume that all segments possess a "length."

For an interesting use of length and the function concept in a development of rational numbers for elementary or junior high students, see the book *Stretchers and Shrinkers* (Harper & Row, 1969) by Peter G. Braunfeld.

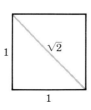 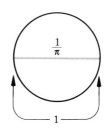 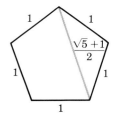

It is possible to describe the length of figures other than segments. For example, we often speak of the length of an arc, or the length of a curve, or the length of a polygonal path. The lengths of these figures can be derived from the following three basic properties of the length function.

Additive Property.   *The length of a figure composed of two segments which have been joined end to end is the sum of the lengths of the two segments.*

$$L(\overline{AB} \cup \overline{BC}) = L(\overline{AB}) + L(\overline{BC})$$

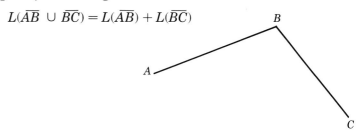

Invariance Property.   *Length does not depend on position. That is, if $\overline{AB}$ is congruent to $\overline{A'B'}$, then*

$$L(\overline{AB}) = L(\overline{A'B'}).$$

Unit Property.   *The length of the unit segment is one.*

EXERCISE SET 11.1

1. The **perimeter** of a polygon is the sum of the lengths of
   the segments of the polygon. Write the simplest formula
   possible for
   a) the perimeter
      of a rectangle;

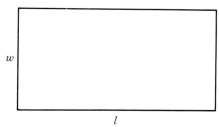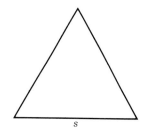

   b) the perimeter of an equilateral triangle;

   c) the perimeter of the regular n-gon with side length l.

2. In Chapter 1, the Pythagorean Theorem is mentioned.
   Recall that this theorem says that if a right triangle has
   hypotenuse with length c and arms of length a and b,
   then $c^2 = a^2 + b^2$. Use this theorem when necessary and
   compute the lengths of the 5 different length segments on
   a geoboard with 3 nails on a side.

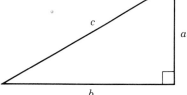

3. Use the Pythagorean Theorem to develop a formula for
   the perimeter of an equilateral triangle which has an altitude with length h. (Recall that an altitude of an equilateral triangle bisects the base.)

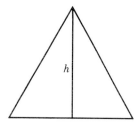

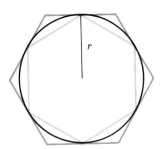

4. Draw a circle; call its radius r. Construct an inscribed regular hexagon, and a circumscribed regular hexagon.

   a) Compute the perimeters of the two hexagons.

   b) How do the perimeters of the hexagons compare with the circumference of the circle?

5. a) Draw circles and use a device of your choice to measure the *circumference* (distance around the circle) as accurately as possible to complete this table.

Radius of ⊙	Diameter d of ⊙	Circumference C of ⊙	$\dfrac{C}{d}$
1			
2			
3			
4			
5			
6			
7			

   b) If you could measure the circumference more precisely, what do you think would be true about the value $\dfrac{C}{d}$ in each case?

6. Use the information in Exercise 5 and the irrational number π (approximately equal to 3.1416) to write a formula for finding the *circumference* C of a circle.

7. Given a segment with length l, can you study this diagram and construct with ruler and compass a segment with length $\dfrac{l}{3}$?

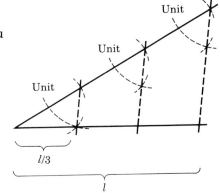

8. The famous Yang-Yin symbol on the right has been constructed from a circle of radius r. Compute the length of the arc which divides the black and white regions.

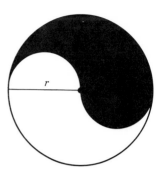

Yang-yin

9. Suppose a stiff red hoop is fitted tightly around the equator of the earth. The diameter of the earth is about 12,800 kilometers. If the hoop is cut and its circumference increased by 10 meters, and the parts replaced so that the hoop is the same distance from the earth at every point, will you be able to

a) put your finger under it?

b) crawl under it?

c) walk under it standing straight up?

d) drive a bus under it?

e) none of these?

Show your calculations to prove your answer.

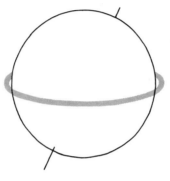

10. a) Compute the perimeter of the figure below.

   b) Explain how you used each of the three properties of length listed on page 364.

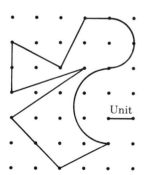

Unit

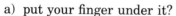

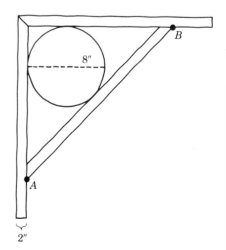

*11. A carpenter wished to place a 45° brace 2 inches thick between two beams so that there is just enough room for a pipe 8 inches in diameter. Find the length of the longer edge AB.

AREA

The geoboard provides an interesting setting in which to set the stage for considering the concept of area measure.

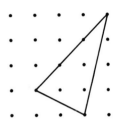

Perhaps one of the most valuable concepts that can be developed on the geoboard is that of area. The book *Geoboard Geometry for Teachers* (Scott Foresman, 1972) by John J. DelGrande has some ideas appropriate for students at various levels.

Investigation 11.2

A. Find the areas for each of the triangles below without using any measuring device and without using any formulas.

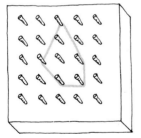

B. The figure on the left has 5 nails touching the rubber band and 2 nails inside. What is its area?

Figures of how many other areas can you construct on a geoboard using one rubber band touching 5 nails with 2 nails inside?

C. Select a counting number n between 3 and 6 and a counting number k between 1 and 5. Construct several different-shaped figures on a geoboard with the band touching n nails with k nails in the interior. Of how many different areas are these figures?

Repeat this process several times.

We see that with each figure constructed in the above investigation we are able to associate a number, called its area. However, for some regions, even some rectangular regions, this number is not always obtainable using a physical measurement process. The following diagrams suggest a physical process that can be abstracted to identify one real number as the measure of region *ABCD*.

Area (*ABCD*) ≈ 18 square units

To find a closer approximation, divide the unit into 100 parts, each 1/100 of the unit

Area (*ABCD*) ≈ 18 + 4.75 ≈ 22.75

As this process is continued, the next smaller unit would be 1/100 of 1/100 of the unit, or 1/10,000 of the original unit. Thus the area of *ABCD* might be:

$$18 + \frac{475}{100} + \frac{2724}{10,000} + \frac{6497}{1,000,000} + \ldots \quad \text{or } 23.028897$$

In this way we generate a decimal for the area of *ABCD*. If the decimal repeats, the area is a rational number. If the decimal is nonrepeating but nonending, the area is an irrational number.

So we see that just as length measure is a function, area measure is a *function* that associates a *unique number* with each *plane region*. To describe the situation pictured on the right, we write *A(WXYZ)* = 8. (We read "the area measure of region *WXYZ* is 8.")

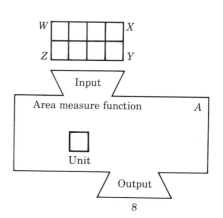

Each plane region has associated with it *exactly one* number, called its area.

The area function satisfies properties which are analogous to the properties satisfied by the length function. The area measure of regions bounded by polygons or simple closed curves can be derived by using these properties.

Additive Property.   *If a region R can be decomposed into a finite number of regions, overlapping only on points, segments, or curves, the area of the entire region is the sum of the areas of the separate regions.*

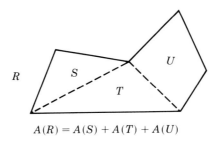

$$A(R) = A(S) + A(T) + A(U)$$

Invariance Property.   *If regions R and S are congruent, then $A(R) = A(S)$.*

Unit Property.   *The area of a unit square is 1.*

All three of the above properties are important when determining the areas for regions bounded by a polygon. It should be interesting for the reader to check to see how he used these properties when completing Investigation 11.2.

There are two additional properties which are of particular value for developing formulas for computing the area of polygon regions. The first describes the relation of area to length and the second describes how regions can be transformed into new regions of equal area.

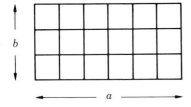

Relation of Area to Length.   *A rectangle with sides of length a units and b units, where a and b are any positive real number, has an area of a × b square units.*

Dissection Property. *If a region can be dissected into parts and these parts reassembled to form another region, then the two regions have equal areas.*

Using this principle, we conclude that the area of Fig. A below is equal to the area of Fig. B.

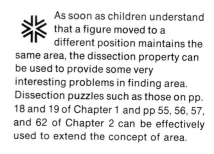

As soon as children understand that a figure moved to a different position maintains the same area, the dissection property can be used to provide some very interesting problems in finding area. Dissection puzzles such as those on pp. 18 and 19 of Chapter 1 and pp 55, 56, 57, and 62 of Chapter 2 can be effectively used to extend the concept of area.

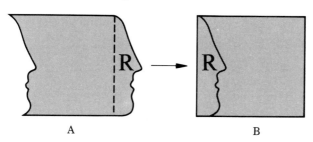

A   B

The following pictures and the directions that accompany them suggest how this principle is used to develop the formulas for finding the area of a parallelogram and a triangle.

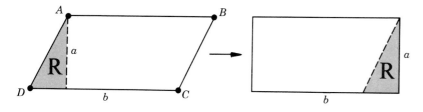

Move region R from the left side to the right side. The area of the parallelogram is the same as the area of the rectangle.

$$A(\text{parallelogram } ABCD) = b \times a.$$

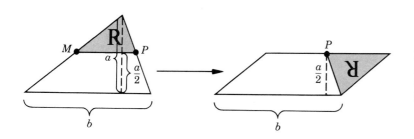

Rotate region R 180° about point P. The area of the triangle is the same as the area of the parallelogram.

$$A \text{ (triangle)} = b \times \frac{a}{2} = \frac{b}{2} \times a$$

$$= \frac{1}{2} \times b \times a.$$

The formulas for finding the areas of other plane regions will be developed in the following exercises.

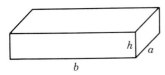

EXERCISE SET 11.2

1. Write a formula for finding the *surface area* of a cuboid if you know lengths b, a, and h. Simplify the formula as much as possible.

2. Write a formula for finding the area of a regular hexagon,
   a) if you know only the lengths d and s,
   b) if you know only the lengths s and r.

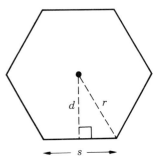

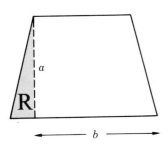

3. Show by dissecting and reassembling region R that the area of an isosceles trapezoid is $b \times a$. You may want to use scissors and construction paper.

4. Every trapezoid can be divided into two triangles, as indicated by the shaded regions I and II in this figure. The altitude (a) of each triangle is the same. Use this information to prove that

   Area of trapezoid $ABCD$ = Area $\triangle$I + Area $\triangle$II

   $$= \frac{1}{2} a(b_1 + b_2)$$

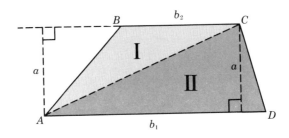

5. a) Draw circles on graph paper and count squares (estimate parts of squares, etc.) to complete the following table.

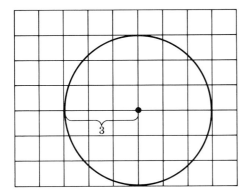

Radius r	n^2	Estimated area of (A)	A/r^2
3			
4			
5			
6			
7			
8			
9			
10			

b) What did you discover in the last column of the table?

6. The area of a circle can be estimated quite accurately by dividing it up into triangles.

When these triangles are reassembled as shown below, the area of the "rectangle" is approximately the area of the circle.

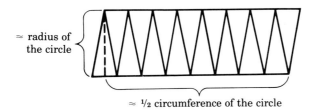

$\approx$ radius of the circle

$\approx \frac{1}{2}$ circumference of the circle

a) Explain how to "develop" the formula for the area of the circle from the "rectangle."

b) What happens if the circle is divided into a greater number of triangles?

7. Find the area of the shaded ring.

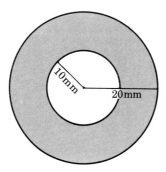

8. a) Draw several ellipses on graph paper using two pins and a thread as shown. Estimate the area of the ellipses and complete this table.

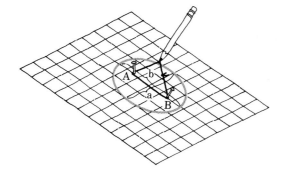

a	b	$a \times b$	Estimated area

b) What did you discover from the table?

9. a) Suppose a rectangle has perimeter 36 units. Show all the possible shapes for the rectangle on graph paper.

b) Which of these shapes has the maximum area?

*10. If X is any point inside an equilateral triangle, the sum of the $\perp$ distances $a + b + c$ is equal to h, the altitude of the triangle. Express the area of the shaded triangles. Use the idea that the sum of these areas is equal to the area of the triangle ABC to prove this theorem.

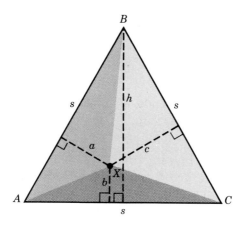

*11. It is possible to construct on a geoboard with 6 nails on a side at least 55 squares of different areas. Find at least 15 of these squares and show them on dot paper. [*Hint:* They cannot all be constructed by using a single rubber band. Two bands are required for some.]

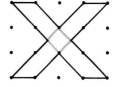

*12. Find the area and perimeter for each polygon below without using any measuring device and without using any formulas (except, perhaps, the pythagorean theorem). Which of the five properties of the area function do you need to use?

*13. A triangle of minimum area on a 5 × 5 geoboard has area
$1/2$ unit. The quadrilateral of minimum area has area 1
unit, and the pentagon of minimum area has area $3/2$
units.

a) Draw on 5 × 5 geopaper 4 different-shaped triangles
each with area $1/2$ unit.

b) Draw at least 8 different-shaped quadrilaterals each
with area 1 unit.

c) Draw at least 8 different-shaped pentagons each with
area $3/2$ units.

What condition relative to the number of nails on the
boundary and in the interior characterizes those polygons
of minimal area?

*14. In Investigation 11.2 we learned that the areas of regions
R_1 and R_2 are equal provided that their boundaries touch
an equal number of nails with an equal number of nails
on the inside of the region.

a) Complete the tables below and search for a formula
which describes the area of a region whose boundary
touches n nails.

Region whose boundary touches n nails with 0 nails in the interior			Region whose boundary touches n nails with 1 nail in the interior		
n	$n/2$	Area of region	n	$n/2$	Area of region
3	$3/2$		3	·	
4	2		4	·	
5	·		5	·	
6	·		6		
·	·		·		
·			·		
·			·		
n			n		

b) Complete tables like those in part (a) for two nails in the interior and three nails in the interior. If n nails are touching the rubber band and p nails are inside, then the *area* of the region enclosed = _____.

*15. The accompanying figure represents a fixed stick AB and a movable stick CD of equal length joined by two rubber bands AB and BC. Suppose CD is moved along XY which is parallel to AB. What effect does this movement have upon the area of parallogram $ABCD$?

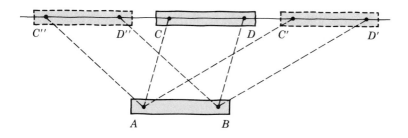

*16. Show that the triangle formed by lines drawn from the midpoint of one of the oblique sides of a trapezoid to the opposite vertices has area $\frac{1}{2}$ the area of the trapezoid.

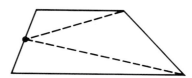

*17. Any point inside a parallelogram is joined to the four vertices. Show that the sum of the areas of either pair of opposite triangles is $\frac{1}{2}$ the area of the parallelogram.

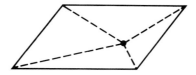

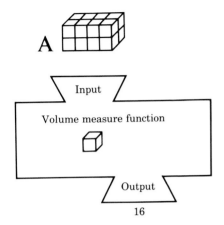

A

Input

Volume measure function

Output

16

VOLUME

Just as for length and area, volume measure may be thought of as a function that associates a unique number with each space figure. To describe the situation pictured here, we write $V(A) = 16$. We read "the volume measure of Figure A is 16."

As with measure of length and area, each space figure has associated with it *exactly* one number. This number is called the volume of the figure and is not always obtainable for a certain figure using a physical process. However, through a choice of smaller and smaller units, each $\frac{1}{1000}$ of the previous unit, the physical process yields approximations closer and closer to the number (either rational or irrational) called the volume of the figure.

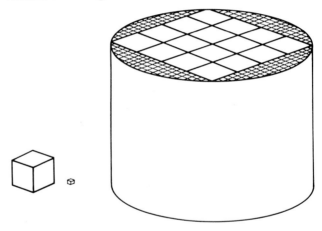

An interesting experiment for children involves the use of sand, rice, crystal sugar, or liquid, and open-top containers as pictured below.

Children are asked how many pyramids-full of the material it will take to fill the prism. They are surprised to find that it takes three and that the volume of a pyramid is one-third the volume of a prism with the same base and height. This idea is also true for cones and cylinders. An experimental approach actually using the materials — the material and the containers — can help the students have a meaningful understanding of this formula.

The volume function satisfies additive, unit, and invariance properties similar to the length and area functions.

Additive Property. *If R and S are objects that have only surfaces in common, then $V(R \cup S) = V(R) + V(S)$.*

Unit Property. *The volume of a unit object is 1.*

Invariance Property. *Volume does not depend on the position of the object in space.*

Two additional properties are important for developing formulas for computing the volume of certain space figures.

One relates volume to length and the second relates volume to cross-sectional area.

Volume and Length. *A box (cuboid) with dimensions l, w, and h has volume l × w × h cubic units.*

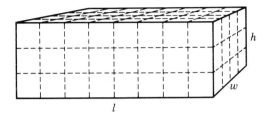

For whole number values of l, w, and h, there are $l \times w$ units in each layer and there are h layers. Hence, the total number of units is found by the formula $V = l \times w \times h$. This formula also applies if the lengths l, w, and h are not whole numbers.

Volume and Cross Sections. *If the area of the base of a space figure with height h is A, and all cross sections of the figure parallel to the base also have area A, then the volume of the figure is A × h.*

This idea is illustrated for four types of solid figures.

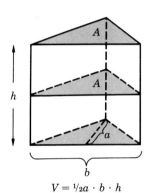

$$V = \tfrac{1}{2}a \cdot b \cdot h$$

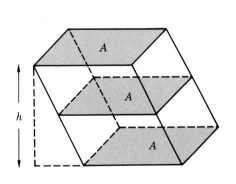

$$V = A \cdot h \qquad\qquad V = A \cdot h$$

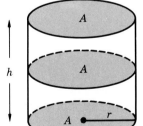

$$V = r^2 h$$

The following exercises will help you develop formulas for finding the volumes of other solid figures.

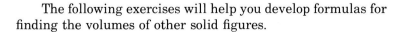

EXERCISE SET 11.3

1. a) Suppose each edge (length, width, height) of a cuboid is doubled. How does this effect the volume (i.e., is the volume double also?)

   b) How is the volume changed if each dimension is multiplied by 10?

   c) Suppose the values of l, w, h are multiplied by a factor k. By what factor is the volume multiplied?

2. a) Find the volume and surface area of this cuboid.

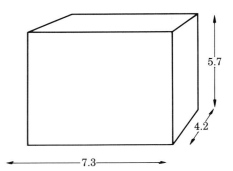

   b) Describe a cuboid with volume equal to surface area.

3. Develop a formula for the volume of

   a) an equilateral triangular prism, and

   b) a regular hexagonal prism.

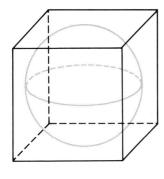

4. Use volume models (if available) and an experimental approach to conjecture formulas for the volume of:

   a) a tetrahedron,

   b) a square pyramid,

   c) a cone,

   d) a hexagonal pyramid.

5. If the volume of a cube is 1, guess at the volume of a sphere inscribed in the cube.

6. The surface area of a sphere is four times the area enclosed by the great circle shown here. Surface area = $4\pi r^2$.

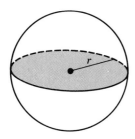 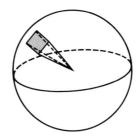

Recall that the volume of a pyramid is $\frac{1}{3}Ah$ (where A is the area of the base). Each square unit of area on the surface of a sphere is the base of a pyramid with approximate height r (radius of the sphere).

a) Accept and use these ideas to develop a formula for the volume of a sphere.

b) Use the formula to check your estimate in Exercise 5.

7. a) Complete this table

Length of edge of cube (cm)	Surface area of cube (cm²)	Volume of cube (cm³)	Ratio of volume to surface area
1			
2			
3			
4			
5			
6			
7			
8			
9			
10			

b) What conclusions can you draw about the ratio of volume to surface area of a cube?

8. What is the ratio of the volume of the sphere to the volume of the touching cylinder?

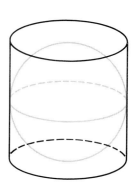

9. One sphere has twice the radius of another. How do the volumes of the spheres compare?

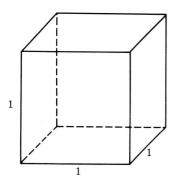

*10. a) If the volume of a unit cube is 1, estimate the volume of each of the regular polyhedra (let all the edges of the polyhedra have length equal to the edge of the cube).

b) Make models and use rice or another material to check your estimates.

*11. Pictured is a stack of 14 cubes. If the volume of one block is one unit, the volume of the stack is 14 units and the surface area is 42 units.

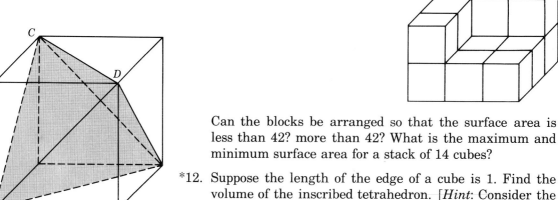

Can the blocks be arranged so that the surface area is less than 42? more than 42? What is the maximum and minimum surface area for a stack of 14 cubes?

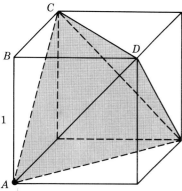

*12. Suppose the length of the edge of a cube is 1. Find the volume of the inscribed tetrahedron. [*Hint*: Consider the volume of pyramid *ABCD*.]

BIBLIOGRAPHIC REFERENCES

[48, 29–53], [57], [58, 294–328], [68, 269–282, 297–309], [77, 1797–1813]

PEDAGOGICAL ACTIVITIES FOR THE TEACHER

1. Write a sequence of discovery exercises which would lead a student to discover the formula for the area of (a) a triangle, (b) a parallelogram, (c) a trapezoid.

2. Choose one of the formulas relating to area or volume in this chapter and devise a teaching aid which would help students understand why this formula gives the correct area or volume. Field test your teaching aid and revise it if necessary.

3. Study the strands of measurement in a selected elementary school mathematics textbook series. Outline the topics presented and the grade level at which they are presented. What aspects of the metric system would you present to students at a given grade level which are not presented in the text?

4. Develop an overhead projectural with overlays which could be used to help students understand the formula for the area of a trapezoid. Use color to help make the ideas stand out.

5. Select a laboratory aid such as the "Sage Kit" (Lapine Scientific Company) which deals with volumes of solid figures or geometry — or "The Circle Kit" (Central Scientific Company) which deals with area and perimeter relationships in the circle. State your criteria and evaluate the appropriateness of these materials for classroom use at a given level.

Appendix A

Compass and Mira Construction

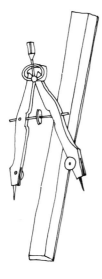

In order to discover relationships in geometry, it is often helpful to draw accurate pictures. The Greek philosopher Plato used only the *compass* and *unmarked straight edge* as the tools for making these pictures. To Plato, a *ruler-compass construction* was performed when "an accurate picture of a geometric idea was produced using only the unmarked ruler and compass according to agreements about the use of these tools." This restriction is historically interesting, and some very important ideas of mathematics have been discovered as people have attempted to make certain constructions using only these tools.

For the purposes of investigating relationships in geometry, however, we will not restrict ourselves to the above tools, but will also often use a protractor and other special tools. One such tool is called a "Mira" and is widely available from producers of commercial aids. The Mira is made from red plexiglass which gives it the property of a mirror and at the same time it is transparent.

The following activities will help you become familiar with the Mira.

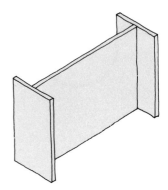

Activity 1

The top of the Mira is different from the bottom of the Mira since the bottom has an angled drawing edge. The angled side should always be toward you.

A. Place the Mira in the correct position and draw a segment along the drawing edge. Label it *PQ*.

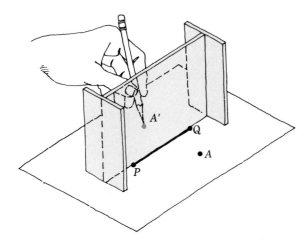

B. Then draw a dot in front of the Mira about 4 cm from the segment. Label it A.

C. Look through the Mira. Do you see a reflection of the dot you just drew?

D. Reach around behind the Mira and mark the image of A on your paper. Label it A'. Draw AA'. Label the point X where AA' intersects PQ.

E. Remove the Mira. Is AA' perpendicular to the original segment? How do the lengths AX and $A'X$ compare?

F. Would the properties in (E) be true for any point like A that you might draw? Try other points.

Activity 2

A. Draw a segment AB on the front side of the Mira.

B. Reach around behind the Mira and draw over the visual image of AB. Label it $A'B'$.

C. Measure the lengths of AB and $A'B'$ to the nearest millimeter. How do the lengths compare?

D. Would your answer in (C) be true for any segment and its image? Try some other segments to be sure.

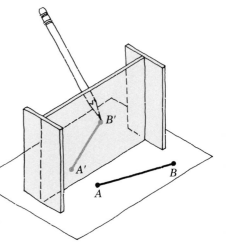

Activity 3

A. Draw a segment along the drawing edge of the Mira (be sure the angled edge is toward you). Label it *AB*. From point *B*, draw a ray *BC* so that an angle is formed as shown.

B. Reach around behind the Mira and mark the visual image of point *C*. Label it *C'*.

C. Remove the Mira and draw ray *BC'*.

D. Use the protractor to measure angles ∠*CBA* and ∠*C'BA*. How do they compare? Try this with other angles.

E. Draw another angle which does not have one side on the drawing edge of the Mira. Draw its image. Does the angle and its image have the same measure?

Activity 4

A. Print your name at the bottom of a sheet of paper in front of the Mira. Reach around behind the Mira and trace the visual image of your name.

B. Cut off your original name so that the visual image remains.

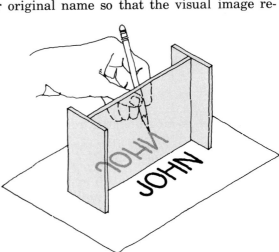

C. Can you place the Mira in relation to the traced image so that your name reappears?

D. Use this idea to write a "mirror message." Give it to someone to decipher.

Activity 5

A. Draw three sides of a square in front of the Mira so that the drawing edge of the Mira serves as the fourth side. Reach around behind the Mira and trace the visual image of the square.

B. Remove the Mira. What type of polygon do you see?

C. Can you draw figures in front of the Mira so that the figure you drew, together with its traced image, forms

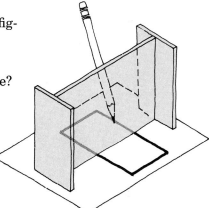

1. a square?
2. a heart?
3. an isosceles triangle?
4. an equilateral triangle?
5. a trapezoid?
6. a regular pentagon?
7. a regular hexagon?
8. a rhombus?
9. an interesting figure?

Activity 6

A. Trace this figure in the center of a sheet of paper.

B. Move your Mira away from the figure. What happens to the image?

C. Move your Mira toward the figure. How does the image move?

D. Place the Mira near the figure and rotate the Mira clockwise (counterclockwise). How does the image move?

E. Can you place the Mira so that a familiar figure is formed by the figure and its image?

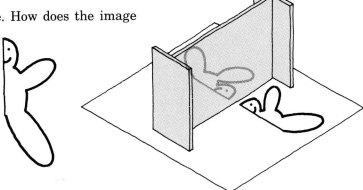

In the following pages we describe how to use the *ruler,* the *compass,* and a *Mira* to perform these basic constructions:

---

1. Copy a circle.

2. Copy a line segment.

3. Copy an angle.

4. Copy a triangle.

5. Construct the perpendicular bisector of a line segment.

6. Bisect an angle.

7. Given a line l and a point P on l, construct a line through P and perpendicular to l.

8. Given a line l and a point P not on l, construct a line through P and perpendicular to l.

9. Given a line l and a point P not on l, construct a line through P and parallel to l.

10. Construct an equilateral triangle, given the length of one side.

Construction 1. Copy a circle.*

A. With a compass

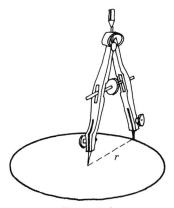

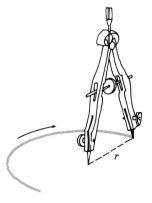

Given circle

1. Measure the radius of the given circle with your compass.

2. Use this radius to draw a copy of the given circle.

---

* Note that a Mira is not very useful for drawing circles. An image of a given circle can be traced freehand if a perfect circle is not required.

B. With a Mira

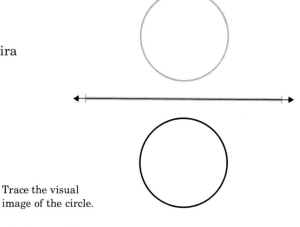

Trace the visual
image of the circle.

Construction 2. Copy a line segment.*

A. With a compass and a ruler

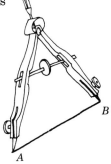

A

Given segment

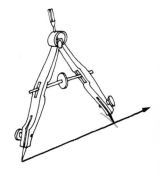

1. Open your compass the length of the given segment.

2. Draw a ray longer than the given segment.

3. Use the same compass to mark a copy of the segment on the ray.

B. With a Mira

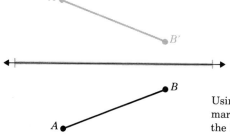

Using the visual image of the segment *AB*, mark points *A'* and *B'*. Use the drawing edge of the Mira to draw the segment *A'B'*.

---

* The possible positions for the copy of a segment, an angle, and a triangle are restricted when using the Mira. A figure and its copy must be the mirror image of each other unless multiple reflections are used.

Construction 3. Copy an angle.

A. With ruler and compass

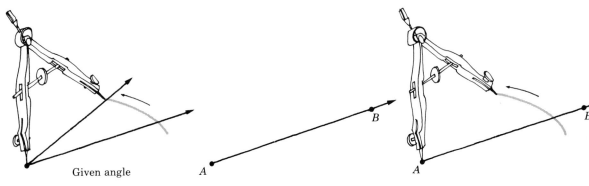

1. Draw an arc intersecting both rays of the given angle.

2. Draw a ray to serve as one side of the copy.

3. With the same compass opening as in (1), draw an arc crossing the ray.

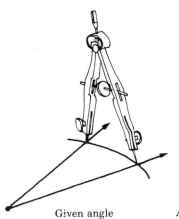

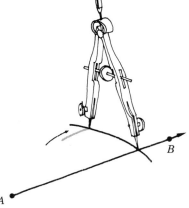

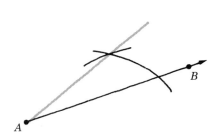

4. Open your compass to measure the opening of the given angle.

5. Use the same opening as in (4) and draw an arc as shown.

6. Draw the second side to complete the copy of the given angle.

B. With a Mira

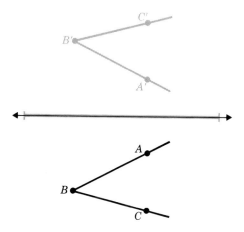

Using the visual image of $\angle ABC$, mark points A', B', and C'. Use the drawing edge of the Mira to complete $\angle A'B'C'$.

Construction 4. Copy a triangle.

A. Using a ruler and compass

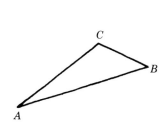

1. Given a triangle, ABC.

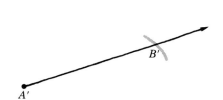

2. Draw a ray and copy segment AB of the triangle.

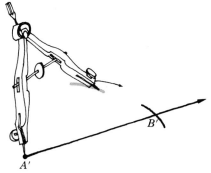

3. Draw an arc with center A and compass opening AC.

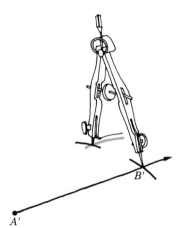

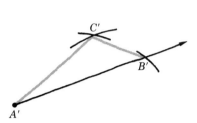

4. Draw an arc with center B and compass opening BC.

5. Use your ruler to complete the drawing of the triangle.

B. Using a Mira

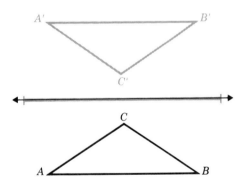

Using the visual image of $\triangle ABC$, mark points A', B', and C'. Use the drawing edge of the Mira to complete the triangle $A'B'C'$. Note that a copy of $A'B'C'$ will be in the same relative position as $\triangle ABC$.

Construction 5. Construct the perpendicular bisector of a line segment.

A. Using a ruler and compass

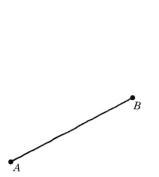

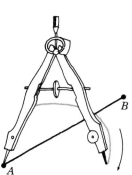

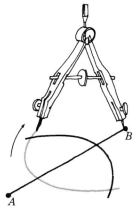

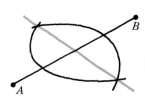

1. Given line segment *AB*.

2. With *A* as center and opening greater than half of *AB*, draw a semi-circular arc.

3. With *B* as center and the same opening as in (2), draw a semicircular arc that intersects the first arc.

4. Connect the two points of intersection to complete the construction of the perpendicular bisector of *AB*.

B. Using a Mira

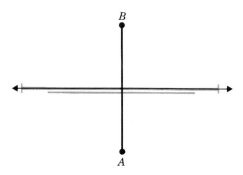

Place the Mira on segment *AB* in such a way that the visual image of point *A* is on point *B*. Draw along the drawing edge of the Mira to produce the perpendicular bisector.

Construction 6. Bisect an angle.

A. Using a ruler and compass

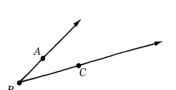

1. Given angle *ABC*.

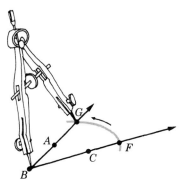

2. With *B* as center, draw an arc that intersects both sides of the angle, at *F* and *G*.

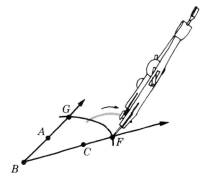

3. With *F* as center, draw an arc in the interior of the angle.

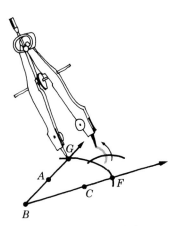

4. With *G* as center, and the same opening as in (3) draw an arc that crosses the first arc.

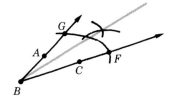

5. Connect *B* and the point of arc intersection to produce the bisector of the angle.

B. Using a Mira

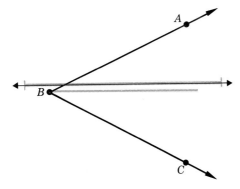

Place the Mira with point B on the drawing
edge in such a way that the visual image of
ray BC is on ray BA. Draw along the drawing
edge of the Mira to produce the angle bisector.

Construction 7. Construct a perpendicular to a line through
a given point on the line.

A. Using a ruler and compass

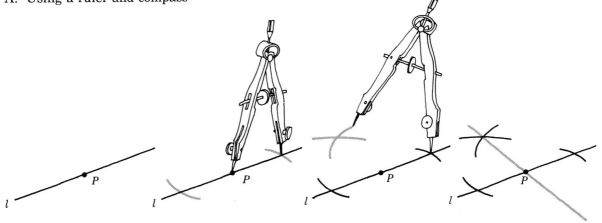

1. Given a line l and a
point P on l.

2. Draw arcs on each
side of P.

3. Draw crossing arcs
above line l.

4. Draw the perpendic-
ular bisector of line l
through P.

B. With a Mira

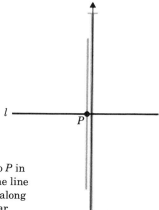

Place the Mira with the drawing edge next to P in such a way that the visual image of half of the line is exactly on the other half of the line. Draw along the drawing edge to produce the perpendicular to line l through point P.

Construction 8. Construct a perpendicular to a line through a given point not on the line.

A. Using a ruler and compass

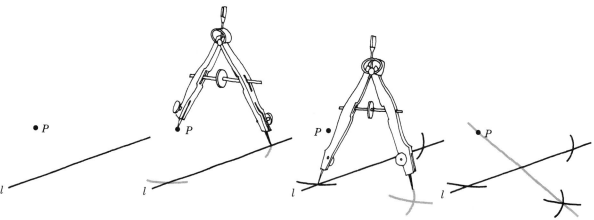

1. Given a line l and a point P not on l.

2. Draw two arcs cutting line l.

3. Draw two crossing arcs below line l.

4. Draw the perpendicular bisector of line l through P.

B. With a Mira

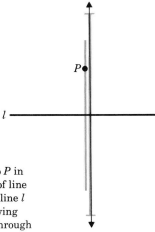

Place the Mira with the drawing edge next to P in such a way that the visual image of the half of line l in front of the Mira is exactly on the half of line l that is behind the Mira. Draw along the drawing edge to produce the perpendicular to line l through point P.

Construction 9. Construct a parallel to a line through a point not on the line.

A. Using a ruler and compass

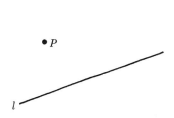

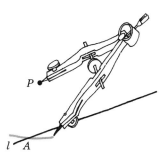

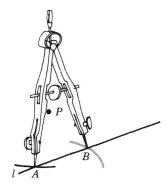

1. Given a line l and a point P not on l.

2. With P as center, draw an arc that crosses line l at the point A.

3. With A as center and the same compass opening, draw an arc that crosses line l at point B.

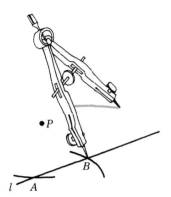

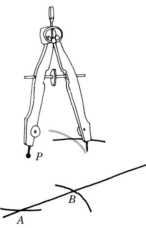

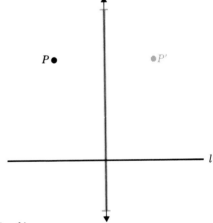

4. With B as center and the same compass opening, draw an arc above line l.

5. With P as center and the same compass opening, draw an arc crossing the arc you drew in (4).

6. Draw the line through P and the intersection of the arcs. This line is parallel to line l.

B. With a Mira

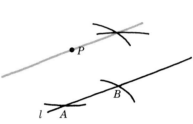

Place one end of the Mira across line l in such a way that the visual image of l falls exactly on l. Then mark the image P' of point P. When PP' is drawn, it will be parallel to line l.

Construction 10. Construct an equilateral triangle, given the length of one side.

A. Using a ruler and compass

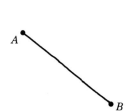

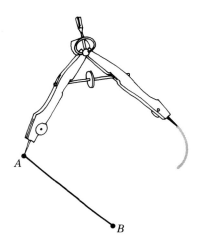

1. Given side *AB* of an equilateral triangle.

2. With *A* as center and compass opening *AB*, draw an arc centered above segment *AB*.

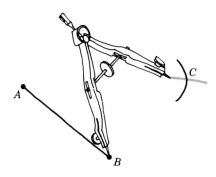

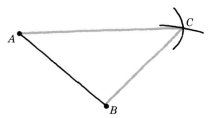

3. With *B* as center and compass opening *BA*, draw an arc that crosses the arc you drew in (2) at point *C*.

4. Draw sides *AC* and *BC* to complete the construction of the equilateral triangle.

B. With a Mira

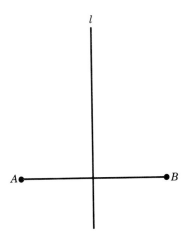

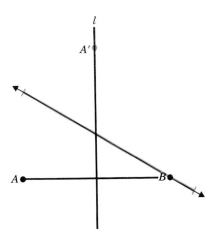

1. Construct the perpendicular bisector l of side AB.

2. Place the drawing edge of the Mira next to B in such a way that the visual image of A is on l. Mark this image A' and connect AA' and BA' to complete the equilateral triangle.

Other more detailed constructions can be performed using combinations of the ten basic constructions just described. Here are a few which you might like to try.

1. Given any three points, construct a circle passing through them. [*Hint*: The perpendicular bisector of a chord of a circle goes through the center of the circle. Use basic construction 5 twice to find the center of the circle.]

2. Construct a tangent to a circle at a given point on the circle [*Hint*: A tangent to a circle is perpendicular to the radius of the circle at the point of tangency. Draw a radius, extend it, and use basic construction 7.]

3. Construct a rhombus, given the length of each of the diagonals of the rhombus [*Hint*: The diagonals of a rhombus are perpendicular. Basic constructions 2 and 5 should help.]

Some references which provide other activities and ideas about ruler and compass and Mira constructions are:

Editors, *Mira Math For the Elementary School*, A Mira Math Co. publication, 1973. (Distributed by Creative Publications.)

Gillespie, N. J. *Mira Activities for Junior High School Geometry,* Mira Math Co., 1973. (Distributed by Creative Publications.)

Appendix B
Postulates and Theorems

The following Postulates and Theorems from plane geometry are provided for your review and reference. They are adaptations of those listed in *Geometry* by Edwin Moise and Floyd Downs (Addison Wesley Publ. Co., Reading, Mass., 1971).

POSTULATES

Postulate 1. *The Distance Postulate. To every pair of different points there corresponds a unique positive number.*

Postulate 2. *The Ruler Postulate. The points of a line can be placed in correspondence with the real numbers in such a way that:*
1. *To every point of the line there corresponds exactly one real number;*
2. *To every real number there corresponds exactly one point of the line; and*
3. *The distance between any two points is the absolute value of the difference of the corresponding numbers.*

Postulate 3. *The Ruler Placement Postulate. Given two points P and Q of a line, the coordinate system can be chosen in such a way that the coordinate of P is zero and the coordinate of Q is positive.*

Postulate 4. *The Line Postulate. For every two points there is exactly one line that contains both points.*

Postulate 5.
a) *Every plane contains at least three noncollinear points.*
b) *Space contains at least four noncoplanar points.*

Postulate 6. *If two points of a line lie in a plane, then the line lies in the same plane.*

Postulate 7. *The Plane Postulate. Any three points lie in at least one plane, and any three noncollinear points lie in exactly one plane.*

Postulate 8. *If two different planes intersect, then their intersection is a line.*

Postulate 9. *The Plane Separation Postulate. Given a line and a plane containing it. The points of the plane that do not lie on the line form two sets such that:*
1. *Each of the sets is convex, and*
2. *If P is in one of the sets and Q is in the other, then the segment $\overline{PQ}$ intersects the plane.*

Postulate 10. *The Space Separation Postulate. The points of space that do not lie in a given plane form two sets, such that:*
1. *Each of the sets is convex, and*
2. *If P is in one of the sets and Q is in the other, then the segment $\overline{PQ}$ intersects the plane.*

Postulate 11. *The Angle Measurement Postulate. To every angle $\angle BAC$ there corresponds a real number between 0 and 180.*

Postulate 12. *The Angle Construction Postulate. Let $\overrightarrow{AB}$ be a ray on the edge of the half-plane H. For every number r between 0 and 180 there is exactly one ray $\overrightarrow{AP}$, with P in H, such that $m\angle PAB = r$.*

Postulate 13. *The Angle Addition Postulate. If D is in the interior of $\angle BAC$, then $m\angle BAC = m\angle BAD + m\angle DAC$.*

Postulate 14. *The Supplement Postulate. If two angles form a linear pair, then they are supplementary.*

Postulate 15. *The SAS Postulate. Every SAS correspondence is a congruence.*

Postulate 16. *The ASA Postulate. Every ASA correspondence is a congruence.*

Postulate 17. *The SSS Postulate. Every SSS correspondence is a congruence.*

Postulate 18. *The Parallel Postulate. Through a given external point there is only one parallel to a given line.*

Postulate 19. *The Area Postulate. To every polygonal region there corresponds a unique positive real number.*

Postulate 20. *The Congruence Postulate. If two triangles are congruent, then the triangular regions determined by them have the same area.*

Postulate 21. *The Area Addition Postulate. Suppose that the region R is the union of two regions R_1 and R_2. Suppose that R_1 and R_2 intersect in at most a finite number of segments and points. Then $aR = aR_1 + aR_2$.*

Postulate 22. *The Unit Postulate. The area of a square region is the square of the length of its edge.*

Postulate 23. *The Unit Postulate. The volume of a rectangular parallelepiped is the product of the altitude and the area of the base.*

Postulate 24. *Cavalieri's Principle. Given two solids and a plane. Suppose that every plane parallel to the given plane, intersecting one of the two solids, also intersects the other, and gives cross sections with the same area. Then the two solids have the same volume.*

THEOREMS

Theorem 1. *Let $\overrightarrow{AB}$ be a ray, and let x be a positive number. Then there is exactly one point P of $\overrightarrow{AB}$ such that $AP = x$.*

Theorem 2. *Every segment has exactly one midpoint.*

Theorem 3. *If two different lines intersect, their intersection contains only one point.*

Theorem 4. *If a line intersects a plane not containing it, then the intersection contains only one point.*

Theorem 5. *Given a line and a point not on the line, there is exactly one plane containing both.*

Theorem 6. *Given two intersecting lines, there is exactly one plane containing both.*

Theorem 7. *If two angles are complementary, then both are acute.*

Theorem 8. *Every angle is congruent to itself.*

Theorem 9. *Any two right angles are congruent.*

Theorem 10. *If two angles are both congruent and supplementary, then each is a right angle.*

Theorem 11. *Supplements of congruent angles are congruent.*

Theorem 12. *Complements of congruent angles are congruent.*

Theorem 13. *Vertical angles are congruent.*

Theorem 14. *If two intersecting lines form one right angle, then they form four right angles.*

Theorem 15. *Every segment is congruent to itself.*

Theorem 16. *Every angle has exactly one bisector.*

Theorem 17. *If two sides of a triangle are congruent, then the angles opposite these sides are congruent.*

Theorem 18. *If two angles of a triangle are congruent, then the sides opposite them are congruent.*

Theorem 19. *In a given plane, through a given point of a given line, there is one and only one line perpendicular to the given line.*

Theorem 20. *The perpendicular bisector of a segment, in a plane, is the set of all points of the plane that are equidistant from the endpoints of the segment.*

Theorem 21. *Through a given external point there is at least one line perpendicular to a given line.*

Theorem 22. *Through a given external point there is at most one line perpendicular to a given line.*

Theorem 23. *If M is between A and C on a line L, then M and A are on the same side of any other line that contains C.*

Theorem 24. *If M is between B and C, and A is any point not on $\overleftrightarrow{BC}$, then M is in the interior of $\angle BAC$.*

Theorem 25. *An exterior angle of a triangle is greater than each of its remote interior angles.*

Theorem 26. *Every SAA correspondence is a congruence.*

Theorem 27. *Given a correspondence between two right triangles. If the hypotenuse and one leg of one of the triangles are congruent to the corresponding parts of the second triangle, then the correspondence is a congruence.*

Theorem 28. *If two sides of a triangle are not congruent, then the angles opposite them are not congruent, and the larger angle is opposite the longer side.*

Theorem 29. *If two angles of a triangle are not congruent, then the sides opposite them are not congruent, and the longer side is opposite the larger angle.*

Theorem 30. *The shortest segment joining a point to a line is the perpendicular segment.*

Theorem 31. *The sum of the length of any two sides of a triangle is greater than the length of the third side.*

Theorem 32. *If two sides of one triangle are congruent, respectively, to two sides of a second triangle, and the included angle of the first triangle is larger than the included angle of the second, then the third side of the first triangle is longer than the third side of the second.*

Theorem 33. *If two sides of one triangle are congruent respectively to two sides of a second triangle, and the third side of the first triangle is longer than the third side of the second, then the included angle of the first triangle is larger than the included angle of the second.*

Theorem 34. *If B and C are equidistant from P and Q, then every point between B and C is equidistant from P and Q.*

Theorem 35. *If a line is perpendicular to each of two intersecting lines at their point of intersection, then it is perpendicular to the plane that contains them.*

Theorem 36. *Through a given point of a given line there passes a plane perpendicular to the given line.*

Theorem 37. *If a line and a plane are perpendicular, then the plane contains every line perpendicular to the given line at its point of intersection with the given plane.*

Theorem 38. *Through a given point of a given line there is only one plane perpendicular to the line.*

Theorem 39. *The perpendicular bisecting plane of a segment is the set of all points equidistant from the end points of the segment.*

Theorem 40. *Two lines perpendicular to the same plane are coplanar.*

Theorem 41. *Through a given point there passes one and only one* plane *perpendicular to a given* line.

Theorem 42. *Through a given point there passes one and only one* line *perpendicular to a given* plane.

Theorem 43. *The shortest segment to a plane from an external point is the perpendicular segment.*

Theorem 44. *Two parallel lines lie in exactly one plane.*

Theorem 45. *Two lines in a plane are parallel if they are both perpendicular to the same line.*

Theorem 46. *Let L be a line and let P be a point not on L. Then there is at least one line through P, parallel to L.*

Theorem 47. *If two lines are cut by a transversal, and one pair of alternate interior angles are congruent, then the other pair of alternate interior angles are also congruent.*

Theorem 48. *Given two lines cut by a transversal. If a pair of alternate interior angles are congruent, then the lines are parallel.*

Theorem 49. *Given two lines cut by a transversal. If a pair of corresponding angles are congruent, then a pair of alternate interior angles are congruent.*

Theorem 50. *Given two lines cut by a transversal. If a pair of corresponding angles are congruent, then the lines are parallel.*

Theorem 51. *If two parallel lines are cut by a transversal, then alternate interior angles are congruent.*

Theorem 52. *If two parallel lines are cut by a transversal, each pair of corresponding angles are congruent.*

Theorem 53. *If two parallel lines are cut by a transversal, the interior angles on the same side of the transversal are supplementary.*

Theorem 54. *In a plane, if two lines are each parallel to a third line, then they are parallel to each other.*

Theorem 55. *In a plane, if a line is perpendicular to one of two parallel lines it is perpendicular to the other.*

Theorem 56. *For every triangle, the sum of the measures of the angles is 180.*

Theorem 57. *Each diagonal separates a parallelogram into two congruent triangles.*

Theorem 58. *In a parallelogram, any two opposite sides are congruent.*

Theorem 59. *In a parallelogram, any two opposite angles are congruent.*

Theorem 60. *In a parallelogram, any two consecutive angles are supplementary.*

Theorem 61. *The diagonals of a parallelogram bisect each other.*

Theorem 62. *Given a quadrilateral in which both pairs of opposite sides are congruent. Then the quadrilateral is a parallelogram.*

Theorem 63. *If two sides of a quadrilateral are parallel and congruent, then the quadrilateral is a parallelogram.*

Theorem 64. *If the diagonals of a quadrilateral bisect each other, then the quadrilateral is a parallelogram.*

Theorem 65. *The segment between the midpoints of two sides of a triangle is parallel to the third side and half as long.*

Theorem 66. *If a parallelogram has one right angle, then it has four right angles, and the parallelogram is a rectangle.*

Theorem 67. *In a rhombus, the diagonals are perpendicular to one another.*

Theorem 68. *If the diagonals of a quadrilateral bisect each other and are perpendicular, then the quadrilateral is a rhombus.*

Theorem 69. *The median to the hypotenuse of a right triangle is half as long as the hypotenuse.*

Theorem 70. *If an acute angle of a right triangle has measure 30, then the opposite side is half as long as the hypotenuse.*

Theorem 71. *If one leg of a right triangle is half as long as the hypotenuse, then the opposite angle has measure 30.*

Theorem 72. *If three parallel lines intercept congruent segments on one transversal T, then they intercept congruent segments on every transversal T' which is parallel to T.*

Theorem 73. *If three parallel lines intercept congruent segments on one transversal, then they intercept congruent segments on any other transversal.*

Theorem 74. *If a plane intersects two parallel planes, then it intersects them in two parallel lines.*

Theorem 75. *If a line is perpendicular to one of two parallel planes it is perpendicular to the other.*

Theorem 76. *Two planes perpendicular to the same line are parallel.*

Theorem 77. *Two lines perpendicular to the same plane are parallel.*

Theorem 78. *Parallel planes are everywhere equidistant.*

Theorem 79. *All plane angles of the same dihedral angle are congruent.*

Theorem 80. *If a line is perpendicular to a plane, then every plane containing the line is perpendicular to the given plane.*

Theorem 81. *If two planes are perpendicular, then any line in one of them, perpendicular to their line of intersection, is perpendicular to the other plane.*

Theorem 82. *If a line and a plane are not perpendicular, then the projection of the line into the plane is a line.*

Theorem 83. *The area of a rectangle is the product of its base and its altitude.*

Theorem 84. *The area of a right triangle is half the product of its legs.*

Theorem 85. *The area of a triangle is half the product of any base and the corresponding altitude.*

Theorem 86. *The area of a trapezoid is half the product of its altitude and the sum of its bases.*

Theorem 87. *The area of a parallelogram is the product of any base and the corresponding altitude.*

Theorem 88. *If two triangles have the same base b and the same altitude h, then they have the same area.*

Theorem 89. *If two triangles have the same altitude h, then the ratio of their areas is equal to the ratio of their bases.*

Theorem 90. *In a right triangle, the square of the hypotenuse is equal to the sum of the squares of the legs.*

Theorem 91. *If the square of one side of a triangle is equal to the sum of the squares of the other two sides, then the triangle is a right triangle, with its right angle opposite the longest side.*

Theorem 92. *In an isosceles right triangle, the hypotenuse is $\sqrt{2}$ times as long as each of the legs.*

Theorem 93. *If the base of an isosceles triangle is $\sqrt{2}$ times as long as each of the two congruent sides, then the angle opposite the base is a right angle.*

Theorem 94. *In a 30-60-90 triangle, the longer leg is $\sqrt{3}/2$ times as long as the hypotenuse.*

Theorem 95. *If a line parallel to one side of a triangle intersects the other two sides in distinct points, then it cuts off segments which are proportional to these sides.*

Theorem 96. *If a line intersects two sides of a triangle, and cuts off segments proportional to these two sides, then it is parallel to the third side.*

Theorem 97. *Given a correspondence between two triangles. If corresponding angles are congruent, then the correspondence is a similarity.*

Theorem 98. *If $\triangle ABC \sim \triangle DEF$, and $\triangle DEF \cong \triangle GHI$, then $\triangle ABC \sim \triangle GHI$.*

Theorem 99. *Given a correspondence between two triangles. If two pairs of corresponding sides are proportional, and the included angles are congruent, then the correspondence is a similarity.*

Theorem 100. *Given a correspondence between two triangles. If corresponding sides are proportional, then the correspondence is a similarity.*

Theorem 101. *In any right triangle, the altitude to the hypote-*

nuse separates the triangle into two triangles which are similar to each other and to the original triangle.

Theorem 102. *Given a right triangle and the altitude to the hypotenuse.*
1. *The altitude is the geometric mean of the segments into which it separates the hypotenuse.*
2. *Each leg is the geometric mean of the hypotenuse and the segment of the hypotenuse adjacent to the leg.*

Theorem 103. *If two triangles are similar, then the ratio of their areas is the square of the ratio of any two corresponding sides.*

Theorem 104. *Two nonvertical lines are parallel if and only if they have the same slope.*

Theorem 105. *Two nonvertical lines are perpendicular if and only if their slopes are negative reciprocals of each other.*

Theorem 106. *The distance between the points (x_1, y_1) and (x_2, y_2) is $\sqrt{(x_2 - x_1)^2 + (y_2 - y_1)^2}$.*

Theorem 107. *Given $P_1 = (x_1, x_2)$ and $P_2 = (y_1, y_2)$. The midpoint of P_1P_2 is the point $P = \left(\dfrac{x_1 + x_2}{2}, \dfrac{y_1 + y_2}{2}\right)$.*

Theorem 108. *If P is between P_1 and P_2, and $\dfrac{P_1P}{PP_2} = r$, then*
$$P = \left(\frac{x_1 + rx_2}{1 + r}, \frac{y_1 + ry_2}{1 + r}\right).$$

Theorem 109. *Let L be a line with slope m, passing through the point (x_1, y_1). Then every point (x, y) of L satisfies the equation $y - y_1 = m(x - x_1)$.*

Theorem 110. *The graph of equation $y - y_1 = m(x - x_1)$ is the line which passes through the point (x_1, y_1) and has slope m.*

Theorem 111. *The graph of the equation $y = mx + b$ is the line which passes through the point $(0, b)$ and has slope m.*

Theorem 112. *The intesection of a sphere with a plane through its center is a circle with the same center and the same radius.*

Theorem 113. *A line perpendicular to a radius at its outer end is tangent to the circle.*

Theorem 114. *Every tangent to a circle is perpendicular to the radius drawn to the point of contact.*

Theorem 115. *The perpendicular from the center of a circle to a chord bisects the chord.*

Theorem 116. *The segment from the center of a circle to the midpoint of a chord is perpendicular to the chord.*

Theorem 117. *In the plane of a circle, the perpendicular bisector of a chord passes through the center.*

Theorem 118. *In the same circle or in congruent circles, chords equidistant from the center are congruent.*

Theorem 119. *In the same circle or in congruent circles, any two congruent chords are equidistant from the center.*

Theorem 120. *If a line intersects the interior of a circle, then it intersects the circle in two and only two points.*

Theorem 121. *A plane perpendicular to a radius at its outer end is tangent to the sphere.*

Theorem 122. *Every tangent plane to a sphere is perpendicular to the radius drawn to the point of contact.*

Theorem 123. *If a plane intersects the interior of a sphere, then the intersection of the plane and the sphere is a circle. The center of this circle is the foot of the perpendicular from the center of the sphere to the plane.*

Theorem 124. *The perpendicular from the center of a sphere to a chord bisects the chord.*

Theorem 125. *The segment from the center of a sphere to the midpoint of a chord is perpendicular to the chord.*

Theorem 126. *If B is a point of $\overset{\frown}{AC}$, then*

$$m\overset{\frown}{ABC} = m\overset{\frown}{AB} + m\overset{\frown}{BC}.$$

Theorem 127. *The measure of an inscribed angle is half the measure of its intercepted arc.*

Theorem 128. *In the same circle or in congruent circles, if two chords are congruent, then so are the corresponding minor arcs.*

Theorem 129. *In the same circle or in congruent circles, if two arcs are congruent, then so are the corresponding chords.*

Theorem 130. *Given an angle with its vertex on a circle, formed by a secant ray and a tangent ray. The measure of the angle is half the measure of the intercepted arc.*

Theorem 131. *The two tangent segments to a circle from a point of the exterior are congruent and determine congruent angles with the segment from the exterior point to the center.*

Theorem 132. *Given a circle C, and a point Q of its exterior. Let L_1 be a secant line through Q, intersecting C in points R and S; and let L_2 be another secant line through Q, intersecting C in points U and T. Then $QR \times QS = QU \times QT$.*

Theorem 133. *Given a tangent segment $\overline{QT}$ to a circle, and a secant line through Q, intersecting the circle in points R and S. Then $QR \times QS = QT^2$.*

Theorem 134. *Let $\overline{RS}$ and $\overline{TU}$ be chords of the same circle, intersecting at Q. Then $QR \times QS = QU \times QT$.*

Theorem 135. *The graph of the equation $(x - a)^2 + (y - b)^2 = r^2$ is the circle with center (a,b) and radius r.*

Theorem 136. *Every circle is the graph of an equation of the form*

$$x^2 + y^2 + Ax + By + C = 0.$$

Theorem 137. *The graph of the equation $x^2 + y^2 + Ax + By + C = 0$ is (1) a circle, (2) a point, or (3) the empty set.*

Theorem 138. *The perpendicular bisectors of the sides of a triangle are concurrent. Their point of concurrency is equidistant from the vertices of the triangle.*

Theorem 139. *The three altitudes of a triangle are always concurrent.*

Theorem 140. *The bisector of an angle, minus its endpoint, is the set of all points of the interior of the angle that are equidistant from the sides.*

Theorem 141. *The angle bisectors of a triangle are concurrent in a point which is equidistant from the three sides.*

Theorem 142. *The medians of every triangle are concurrent. And their point of concurrency is two-thirds of the way along each median, from the vertex to the opposite side.*

Theorem 143. *Given two circles of radius a and b, with c as the distance between their centers. If each of the numbers a, b, and c is less than the sum of the other two, then the circles intersect in two points, on opposite sides of the line through the centers.*

Theorem 144. *The ratio of the circumference to the diameter is the same for all circles.*

Theorem 145. *The area of a circle of radius r is πr^2.*

Theorem 146. *If two arcs have equal radii, then their lengths are proportional to their measures.*

Theorem 147. *If an arc has measure q and radius r, then its length is*

$$L = \frac{q}{180} \cdot \pi r.$$

Theorem 148. *The area of a sector is half the product of its radius and the length of its arc.*

Theorem 149. *If a sector has radius r and its arc has measure q, then its area is*

$$A = \frac{q}{360} \cdot \pi r^2.$$

Theorem 150. *All cross sections of a triangular prism are congruent to the base.*

Theorem 151. *All cross sections of a prism have the same area.*

Theorem 152. *The lateral faces of a prism are parallelogram regions.*

Theorem 153. *Every cross section of a triangular pyramid, between the base and the vertex, is a triangular region similar to the base. If h is the altitude, and k is the distance from the vertex* to the cross section, then the area of the cross section is equal to k^2/h^2 times the area of the base.

Theorem 154.   *In any pyramid, the ratio of the area of a cross section to the area of the base is k^2/h^2, where h is the altitude of the pyramid and k is the distance from the vertex to the plane of the cross section.*

Theorem 155.   *If two pyramids have the same base area and the same altitude, then cross sections equidistant from the vertices have the same area.*

Theorem 156.   *The volume of any prism is the product of the altitude and the area of the base.*

Theorem 157.   *If two pyramids have the same altitude and the same base area, and their bases lie in the same plane, then they have the same volume.*

Theorem 158.   *The volume of a triangular pyramid is one-third the product of its altitude and its base area.*

Theorem 159.   *The volume of a pyramid is one-third the product of its altitude and its base area.*

Theorem 160.   *Every cross section of a circular cylinder is a circular region congruent to the base.*

Theorem 161.   *Every cross section of a circular cylinder has the same area as the base.*

Theorem 162.   *Given a cone of altitude h, and a cross section made by a plane at a distance k from the vertex. The area of the cross section is equal to k^2/h^2 times the area of the base.*

Theorem 163.   *The volume of a circular cylinder is the product of its altitude and the area of its base.*

Theorem 164.   *The volume of a circular cone is one-third the product of its altitude and the area of its base.*

Theorem 165.   *The volume of a sphere of radius r is $\frac{4}{3}\pi r^2$.*

Theorem 166.   *The surface area of a sphere of radius r is $A = 4\pi r^2$.*

Hints and Answers to Selected Exercises

CHAPTER 1

Exercise Set 1.1

1. Geometry comes into play as one investigates packing problems—from molecules to corn kernels on a cob. Fascinating geometrical problems arise in the study of soap bubbles, tree growth, water flow, and many other topics. The book *Patterns in Nature* by Peter S. Stevens (Boston: Little, Brown, 1974) illustrates many of these ideas.

3. This ratio is 1.623 (to 3 decimal places). The golden ratio is 1.618 (to 3 decimal places). They differ by .005.

6. All "lines" through point P intersect line l.

8.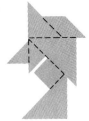

CHAPTER 2

Exercise Set 2.1

1. a) Segment; c) Circle; e) Angle;
   g) Perpendicular lines; i) A plane

2. a) Tip of a needle

   c) Surface of a basketball court

   e) The edges of the top and side of a picture frame

   g) The cutting edges of an opened pair of scissors

   i) Spokes on a bicycle wheel

7. Are the points of intersection collinear? Try several different cases.

9. Are the points A, B, and C collinear? Will this always happen?

Exercise Set 2.2

3. All triangles described in 1 and 2 can be drawn on isometric dot paper. All triangles described in 1 and 2, *except* equilateral triangles, can be drawn on square dot paper.

4. c) No such triangle exists. f) An isosceles triangle

5. Figure in part (e) has 3 lines of symmetry and 120° and 240° rotational or turning symmetry.

9. a) A rectangle with a "V-shaped" piece removed from the center of one side has one line of symmetry, but no rotational symmetry.

   b) Modify the parallelogram to devise such a figure.

11. Formula is $S = (n - 2) \cdot 180$.

14. One approach would be to cut off the corners of a triangular shape and paste them around a point. If a straight line results, the angles have sum of degree measures equal to 180.

   Can you draw a line through the vertex of $\angle 1$ and parallel to the side opposite $\angle 1$ and use the angles associated with these parallel lines to convince someone that $m(\angle 1) + m(\angle 2) + m(\angle 3) = 180$?

Exercise Set 2.3

1. 6

3. a) Heptagon: $51\frac{3}{7}°$; dodecagon: 30°; nonagon: 40°

   b) The polygon has a large number of sides and large vertex angles.

7. a) Regular pentagon: 5 lines; regular hexagon: 6 lines; regular octagon: 8 lines

   b) A regular polygon of n sides has n lines of symmetry.

11. How do the indicated ratios compare to the Golden Ratio (1.618, to the nearest thousandth)?

Exercise Set 2.4

1. For the star polygon $\left\{ \begin{matrix} 7 \\ 3 \end{matrix} \right\}$:

   a) 7 sides   b) V_1V_4, V_4V_7, V_7V_3, V_3V_6, V_6V_2, V_2V_5, and V_5V_1

    c) 7 lines of symmetry

    d) $51^{3/7}°$, $102^{6/7}°$, $154^{2/7}°$, $205^{5/7}°$, $257^{1/7}°$, and $308^{4/7}°$ rotational symmetry.

3. There is exactly one 10-sided regular star polygon.

5. n and d are relatively prime, i.e., the greatest common divisor of n and d is 1.

7. There are four 11-sided star polygons.

9. a) Have you found a trapezoid? an isosceles trapezoid? a rhombus? a kite? or any nonconvex polygons? There are many other geometric figures that can be found.

Exercise Set 2.5

1. The figure should be a parallelogram. Do you get the same result when you start with different shape quadrilaterals? Suppose you divide the sides into thirds and connect the "two-thirds points" rather than the midpoints. What figure do you get?

3. The resulting figure should be an equilateral triangle. What if the triangles were constructed internally? Would you get the same result?

5. All triangles should be congruent.

    b) The resulting triangles should be similar.

7. The segments should be perpendicular. Do you get the same result if you construct the squares internally? What happens if you alternate internal and external constructions?

9. The resulting figure should be a parallelogram. What happens if the triangles are constructed alternately internally and externally in the opposite order?

11. a) Do the points fall on a circle?

    b) The center of the nine point circle lies on the Euler line, midway between the orthocenter and the circumcenter.

CHAPTER 3

Exercise Set 3.1

3. Tessellations in both parts (a) and (b) can be conveniently drawn on equilateral triangular dot paper.

6. a) Vertical angles are congruent.

   b) The segment with endpoints which are midpoints of two sides of a triangle is parallel to the third side and half as long.

   c) The sum of angle measures of a triangle is 180°.

   d) If two triangles are similar, then the lengths of corresponding sides are proportional.

9. a) The dual tessellation of the regular tessellation of squares is again a tessellation of squares.

   b) The dual of the tessellation of equilateral triangles is the tessellation of regular hexagons.

11. a) All twelve pentominoes tessellate the plane.

Exercise Set 3.2

1. $(4, 8, 8)$, $(4, 6, 12)$, $(3, 12, 12)$, $(3, 4, 6, 4)$, $(3, 3, 3, 4, 4)$, $(3, 3, 4, 3, 4)$, $(3, 3, 3, 3, 6)$

CHAPTER 4

Exercise Set 4.1

1. The four match sticks should be placed along the edges of a tetrahedron.

3. b) Yes. One with vertices A, C, F, H.

5. $V + F = E + 2$.

9.   dual
   tetrahedron $\longleftrightarrow$ tetrahedron
   cube $\longleftrightarrow$ octahedron
   octahedron $\longleftrightarrow$ cube
   dodecahedron $\longleftrightarrow$ icosahedron
   icosahedron $\longleftrightarrow$ dodecahedron

11. a) Octahedron, icosahedron

Exercise Set 4.2

3. a) Six, one for each of the six pairs of opposite edges.

   b) Three, one of each pair of opposite faces.

5. a) Four.

   b) Six, if the pentagon base is a regular pentagon.

   c) A cuboid, sometimes called a rectangular parallelepiped, in general has only three planes of symmetry.

   d) An infinite number.

   e) Nine, if octagon is a regular octagon.

   f) Four.

7. a) No. b) No. c) A few of the possible shapes are equilateral triangles, isosceles triangular, scalene triangular, square, rectangle, regular hexagon

9. Twenty-four. The cube may be placed in the box with no rotations performed; each of the 3 axes of order 4 yields 3 other ways the cube may be placed in the box, one through a 90° turn, one by a 180° turn, and one by a 270° turn; each of the 4 axes of order 3 yields 2 other ways; each of the 6 axes of order 2 yields one other way. Summarizing, the axes of order 4 yield 9 ways, the axes of order 3 yield 8 ways, the axes of order 2 yield 6 ways, and one with no rotation, or 24 ways in all.

Exercise Set 4.3

1. a) There is one axis of rotational symmetry of order three. This axis is the line which passes through the vertex that has been truncated and its opposite vertex.

   There are three edges emanating from the vertex which has been truncated. For each of these edges, the plane through the edge and its opposite edge is a plane of symmetry. So there are three planes of symmetry.

   b) It depends on which two vertices are truncated. There are three cases to consider: 1) opposite vertices of the cube, 2) adjacent vertices (joined by an edge), and 3) vertices which are opposite vertices of one of the square faces.

3. Seven of the fourteen semiregular polyhedra are listed:

   Truncated tetrahedron: $(3, 6, 6)$
   Truncated cube: $(3, 8, 8)$

Truncated octahedron: $(4, 6, 6)$
Truncated dodecahedron: $(3, 10, 10)$
Truncated icosahedron: $(5, 6, 6)$
Truncated cube octahedron: $(4, 6, 8)$
Snub cube: $(3, 3, 3, 3, 4)$

CHAPTER 5

Exercise Set 5.2

1. a) Don't cry "foul" — rather cry "fowl"!

    b) Sometimes it's easy to read when you're standing on your head!

    c) Now you see it, now you don't!

    d) A numeral is a name for a number?

5. $(-4, 2)$

7. The image of line p goes through $(1, 10)$ and $(6, 10)$. The image of line q goes through $(2, -1)$ and $(4, -2)$.

Exercise Set 5.3

1. a) Each letter in "CHOICE" remains the same when reflected in the line. This is not true about the letters in "QUALITY."

    b) The image of the letters of "TIMOTHY" upon reflection in the line is the same as the original letter. This is not so for the word "REBECCA."

    c) The mirror image of the "numerals" is

    $$\begin{array}{r} \text{nine} \\ \text{one} \\ \underline{\text{eight}} \\ \text{eighteen} \end{array}$$

5. Yes

9. a) A, H, I, M, O, T, U, V, W, X, Y

    b) Examples are MOM, YAM, MAT. There are others.

11. The reason this looks so strange is that it was written using a mirror.

Exercise Set 5.4

1. a) It is a slide in the direction from p to q (perpendicular to lines p and q) through a distance of 8 centimeters.

3. The result is a half-turn about the point of intersection of lines t and u.

5. n must be one-half the distance of AA' to the right of line m.

7. A half-turn about point C', such that $CC' = \frac{1}{2}AB$ and $CC' \| AB$.

Exercise Set 5.5

1. a) The distance between the lines must be one-half the distance between corresponding points on the two triangles.

   b) Any pair of lines satisfying the condition in (a) and perpendicular to a line segment connecting corresponding points on the two triangles will suffice.

3. The first final position is the same distance to the right of $\triangle ABC$ as the second final position is to the left of $\triangle ABC$. If $p = q$, then the result is the same.

5. a) Not true. b) True. c) True.

CHAPTER 6

Exercise Set 6.1

1. a) $A = (-1, 2)$, $B = (2, 3)$, $C = (3, 1)$, $D = (2, 0)$
   b) $A' = (1, 1)$, $B' = (4, 2)$, $C' = (5, 0)$, $D' = (4, -1)$

3. $C = (3, 2)$

Exercise Set 6.2

3. b) $-60°$

5. a) $A' = (-1, 2)$, $B' = (-2, 5)$, $C' = (-6, 3)$. b) $(-y, x)$

Exercise Set 6.3

1. b) $M_x(A) = (1, -5)$, $M_y(A) = (-1, 5)$. c) $M_x(P) = (x, -y)$, $M_y(P) = (-x, y)$

5. All points on line p.

Exercise Set 6.4

1. a) One is a reflection of the other. c) One is a 180° rotation of the other. d) One is a translation image of the other.

3. Translations and rotations are direct. Reflections and glide reflections are opposite.

Exercise Set 6.6

1. a) $M_\alpha(A) = A$, $M_\alpha(C) = E$, $M_\alpha(D) = D$.
   b) $R_{L,72}(A) = A$, $R_{L,72}(B) = C$.

3. A given axis of symmetry may be the intersection of more than one pair of planes of symmetry.

Exercise Set 6.7

1. Seven

3. The sixth frieze pattern from the top in Fig. 6.25.

CHAPTER 7

Exercise Set 7.1

3. a) $k = 3/2$; b) $k' = 2/3$; and $k' = 1/k$.

5. They are equal.

Exercise Set 7.2

3. *Hint: C'* must lie on the line through A' which is parallel to line AC.

5. A magnification preserves parallelism and the property of being a quadrilateral.

7. Ratios of lengths of segments are preserved by magnifications.

Exercise Set 7.3

3. Always.

Exercise Set 7.4

1. b) $A \leftrightarrow X, B \leftrightarrow Y, C \leftrightarrow Z, D \leftrightarrow W$.

3. Fifty-three feet.

7. a) No. b) Yes.

CHAPTER 8

Exercise Set 8.1

3. The array of numbers below correctly completes the table.

2	2	4	4
1	1	2	2
2	2	4	4
1	1	2	2

5. If you begin with an odd number of half twists, you end up with one band after cutting. If you begin with an even number of half twists, you end up with two bands after cutting.

Exercise Set 8.2

1. a) The third and fourth. c) The third.

5. a) Middle figure. c) Middle figure. d) Right figure.

7. All of them.

Exercise Set 8.3

1. Using the network in Fig. 8.9, we see that there are more than two odd vertices. Thus there is no continuous path which crosses each bridge exactly once.

3. Type II add one, two, two, respectively.
   Type I add two, three, three, respectively.

CHAPTER 9

Exercise Set 9.1

1. a) 2500, 10,000. b) 1275, 5050.

3. a) 4, 6, 8, 10, 12, 14, 15, 16, 18, 20, 21, 22, 24, 26, 27, 28, 30, 32, 33, 34, 35

   b) A prime number is a number greater than 1 that is neither a square nor a rectangular number.

5. $10 + 15 = 5^2$ What can you say about the
    $15 + 21 = 6^2$ sum of two consecutive
    $21 + 28 = 7^2$ triangular numbers?

7. 35, 51, 70

Exercise Set 9.2

1. Number of diagonals = number of sides − 3.

3. Two more pairs in the table are $(4, 12)$ and $(5, 20)$.

Exercise Set 9.3

1. The next 3 pairs in the table are $(3, 7)$, $(4, 11)$, and $(5, 16)$.

2. c) One segment on the 6 dot by 6 dot paper has the same length (5 units) as the segment along one side of the grid. Thus a duplicate is found and the entry in the table is not as expected.

Exercise Set 9.4

1. The next 5 rows are

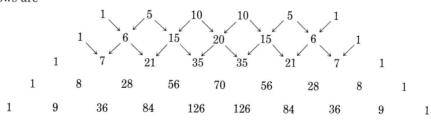

   Note the arrows to see the relationship mentioned.

3. a) Additional pairs in the table are $(3, 4)$, $(4, 8)$, $(5, 16)$, and $(6, 32)$
   b) Yes, $2^{10-1} = 2^9 = 512$. c) $S = 2^{n-1}$ where n is the row number.

5. a) 8th. b) Any row in the sequence 1, 2, 4, 8, 16, 32, etc. contains only odd numbers.

7. a) 2, 3, 7 b) Consider prime numbers.

9. a) 1, 1, 2, 3, 5, 8, 13, 21, 34, 55, 89, 144, 233, 377, 610

   b) The next 3 pairs in the table are (5, 20), (6, 33), and (7, 54).

CHAPTER 10

Exercise Set 10.1

1. a) Devices range from a simple ruler to an odometer to the most complicated electronic sonar equipment and beyond. There are many possibilities.

   b) Try to think of units of length, area, volume, weight, capacity, time, and temperature.

3. A tire pump, for example, could be used to measure "expansion size" of a balloon. One balloon, for example, might be able to stand "10 pumps" before bursting while another might only stand "5 pumps." Can you find other examples?

5. Sample decisions might deal with height, if one is estimating which center will get the tip in basketball, yardage determinations in football, weight and size of bats in baseball, and so on. Also, don't forget problems of refreshments, scoring, admission etc.

Exercise Set 10.2

3. $1/1000$ of one dollar, a penny, a dime, a ten dollar bill, a thousand dollar bill, one dollar.

Exercise Set 10.3

1. a) Thickness of a dime. c) Thickness of a small drinking straw. e) Width of a nickel. g) Height of a small juice glass. i) Height of a door. k) About 4 city blocks.

7. a) About 1500 km. c) About 2400 km. d) About 1300 km. f) About 380 km.

9. a) 8.67. c) 63200. d) 367. f) 64.7.

11. a) Lay string along curve, then measure it.

b) Draw a straight line, mark off small distances on curve and transfer them to the straight line. Traverse the complete curve in this manner.

c) Roll the nickel along the curve, count the number of complete revolutions. Reproduce these on a straight line and measure.

Exercise Set 10.4

1. Square would measure 10 meters on each side.

3. a) A football field is 160 ft by 300 ft, about 5333 sq yds, or about 44 ares.

   c) A 2000 sq ft house would have floor space of 22.2 sq yd or about ⅕ of an are.

5. 6400

9. a) 64 square units. b) 65 square units. c) It can be proved that a thin piece of area 1 square unit is missing along the diagonal when the pieces are assembled to make the rectangle.

Exercise Set 10.5

1. b) A 1-quart milk carton has volume slightly less than 1 cubic decimeter.

3. a) About 250 ml. c) About 3.8 liters. e) About 5 ml.
   g) About 38 liters (a 10-gal tank). i) About 15 ml.

5. a) 1,000,000. c) 1,000,000,000. e) 1,000,000. g) 1,000,000.

7. a) 94.7. c) .947. e) 649.6.

9. a) 6. c) 8. e) 1.

Exercise Set 10.6

1. a) An encyclopedia about 2½−3 cm thick might weigh about 1 kilogram.

3. a) 2 grams. e) About 450 grams. k) About 20 grams. l) About 10 grams. p) About 450 grams.

5. 1.87 kilograms is 1870 grams.

Exercise Set 10.7

1. a) 37°C. b) about 10°C. d) about 80°C. f) 100°C. h) about 15°C.

3. a) 1000. c) 1000. e) 1. g) .001.

7. a) A procedure would be
   i) change 7 ft 8½ inches to inches;
   ii) cube the number of inches;
   iii) divide the answer in (ii) by 1728 (the number of cubic inches in a cubic foot) to arrive at the volume of the tank in cubic feet (ans.: 458.02 cu ft);
   iv) multiply the result in (iii) by 7.48 (the number of gallons in a cubic foot) to find the volume of the tank in gallons (ans.: 3426 gal);
   v) multiply the answer in (iii) by 62.425 (the weight in pounds of a cubic foot of water) to find the weight of the water in pounds (ans.: 28,592 lb)

   b) A procedure would be
   i) cube 3.08 to find the volume in cubic meters (ans.: 29.22 cu m);
   ii) multiply the answer in (i) by 1000 (since there are 1000 cubic decimeters or liters in a cubic meter) to get the volume in liters (ans.: 29,220 liters);
   iii) since 1 liter weighs 1 kilogram, the weight in kilograms is the same as the answer in (ii).

9. a) Use the data that 1 acre = 4840 square yards. (ans.: 51.65).

   b) Use the data that 1 hectare is 10,000 square meters. (ans.: 25 hectares)

CHAPTER 11

Exercise Set 11.1

1. a) $P = 2 (l + w)$
   c) $P = nl$

3. $P = 2\sqrt{3}h$

5. b) The value of $\frac{c}{d}$ should be reasonably close to 3.14, an approximation to π.

9. A person shorter than 5 ft 3 in. will be able to walk under it. Since the diameter (d) is circumference (C) ÷ π, we can set up the following equation.

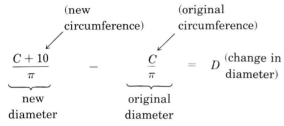

This simplifies to $D = \dfrac{10}{\pi} = 3.18$ meters.

The increase at a given point on the earth would be $\dfrac{3.18}{2}$ or 1.59 meters.

This is about 5 feet 3 inches.

Exercise Set 11.2

2. b) $A = \dfrac{3}{2}s\ \sqrt{4r^2 - s^2}$

5. b) The ratio $\dfrac{A}{r^2}$ should be close to π.

7. 300π square millimeters

9. a) There are 9 possible shapes if the sides of the rectangle are to be whole numbers.

   b) The 9 by 9 square has the maximum area.

Exercise Set 11.3

1. a) The volume is multiplied by 8.

   c) k^3

3. a) $V =$ area of base × height. If the side of the equilateral triangle is s, the formula is $V = \sqrt{3}s^2h/4$.

5. Volume of sphere is $\dfrac{\pi}{6}$ or approximately .52.

7. b) Ratio is $r/3$.

9. One volume is 8 times the other.

Bibliography

1. E. A. Abbott. *Flatland: A Romance in Many Dimensions.* 5th rev. ed., New York: Barnes and Noble Books, 1963.

2. Janet S. Abbott. *Learn to Fold — Fold to Learn.* Pasadena, Calif.: Franklin Publications, 1968.

3. Janet S. Abbott. *Mirror Magic.* Pasadena, Calif.: Franklin Publications, 1968.

4. Irving Adler. *A New Look at Geometry.* New York: Signet Science Library Books, 1966.

5. Association of Teachers of Mathematics. *Mathematical Reflections.* New York: Cambridge University Press, 1970.

6. Association of Teachers of Mathematics. *Notes on Mathematics in Primary Schools.* New York: Cambridge University Press, 1968.

7. P. G. Ball. "A Different Order of Reptile," *Mathematics Teaching,* September 1972, **60:** 44–45.

8. W. W. R. Ball and H. S. M. Coxeter. *Mathematical Recreations and Essays.* New York: The Macmillan Company, 1967.

9. H. V. Baravalle. "The Geometry of the Pentagon and the Golden Section," *The Mathematics Teacher,* January 1948, **41:** 22–23.

10. Stephen Barr. *Experiments in Topology.* New York: Thomas Y. Crowell, 1964.

11. Fred Bassetti, Hy Ruchlis, and Daniel Malament. *Solid Shapes Lab.* New York: Science Materials Center, 1961.

12. Fred Bassetti, Hy Ruchlis, and Daniel Malament. *Math Projects : Polyhedral Shapes.* Brooklyn, New York: Book-Lab, 1968.

13. Robert S. Beard. *Patterns in Space.* Palo Alto, Calif.: Creative Publications, 1973.

14. Anatole Beck, Michael N. Bleicher, and Donald W. Crowe. *Excursions in Mathematics.* New York: Worth Publishers, 1969.

15. David Bergamini. *Mathematics.* Life Science Library, New York: Time-Life Books, 1970.

16. Phil Boorman. "More About Tessellating Hexagons (2)," *Mathematics Teaching,* Summer 1971, **55:** 23.

17. (Edited by:) M. Brydegaard and J. E. Inskeep, Jr. *Readings in Geometry from the Arithmetic Teacher.* Washington, D. C.: National Council of Teachers of Mathematics, 1970.

18. Louis G. Brandes. *Geometry Can Be Fun.* Portland, Me.: J. Weston Walch, 1958.

19. F. J. Budden. *The Fascination of Groups.* London: Cambridge University Press, 1962.

20. J. H. Caldwell. *Topics in Recreational Mathematics.* New York: Cambridge University Press, 1966.

21. A. J. Cameron. *Mathematical Enterprises for Schools.* New York: Pergamon Press, 1966.

22. Stanley Clemens. "Tessellations of Pentagons," *Mathematics Teacher* 67(June 1974).

23. Donald Cohen. *Inquiry in Mathematics Via the Geoboard,* Teacher Guide. New York: Walker Educational Book Corporation, 1967.

24. Richard Courant and Herbert Robbins. *What is Mathematics?* New York: Oxford University Press, 1941.

25. H. S. M. Coxeter. "The Four Color Map Problem, 1890–1940," *The Mathematics Teacher,* April 1959, pp. 283–9.

26. H. S. M. Coxeter. *Introduction to Geometry.* New York: John Wiley & Sons, 1961.

27. H. S. M. Coxeter. *Regular Polytopes.* London: Methuen, 1948.

28. H. S. M. Coxeter and S. L. Greitzer. *Geometry Revisited.* New York: Random House, The L. W. Singer Co., 1967.

29. H. M. Cundy and A. P. Rollett. *Mathematical Models.* London: Oxford University Press, 1961.

30. John Del Grande. *Geoboards and Motion Geometry for Elementary Teachers.* Glenview, Ill.: Scott Foresman, 1972.

31. Daniel V. DeSimone. *A Metric America: A Decision Whose Time Has Come.* Washington, D. C.: National Bureau of Standards, July 1971.

32. James A. Dunn. "More About Tessellating Hexagons (1)," *Mathematics Teaching,* Summer 1971, **55:** 23–24.

33. A. Ehrenfeucht. *The Cube Made Interesting.* New York: A Pergamon Press Book, Macmillan, 1964.

34. H. A. Elliott, James R. MacLean, and Janet M. Jordan. *Geometry in the Classroom—New Concepts and Methods.* Canada: Holt, Rinehart and Winston of Canada, Limited, 1968.

35. M. C. Escher. *The Graphic Work of M. C. Escher.* New York: Hawthorne Publishing, 1960.

36. Howard Eves. *A Survey of Geometry,* Revised Edition. Boston, Mass.: Allyn and Bacon, 1972.

37. David S. Fielker. *Topics from Mathematics — Cubes.* New York: Cambridge University Press, 1969.

38. John N. Fujii. *Puzzles and Graphs.* Washington, D. C.: National Council of Teachers of Mathematics, 1966.

39. Martin Gardner. *The Ambidextrous Universe.* New York: Basic Books, 1964.

40. Martin Gardner. *New Mathematical Diversions from Scientific American.* New York: Simon and Schuster, 1966.

41. Martin Gardner. *The Scientific American Book of Mathematical Puzzles and Diversions.* New York: Simon and Schuster, 1959.

42. Martin Gardner. *The Second Scientific American Book of Mathematical Puzzles and Diversions.* New York: Simon and Schuster, 1961.

43. Martin Gardner. *The Unexpected Hanging.* New York: Simon and Schuster, 1969.

44. Matelo Ghyka. *The Geometry of Art and Life.* New York: Sheed and Ward, 1946.

45. Richard A. Gibbs. "Euler, Pascal, and the Missing Region." *The Mathematics Teacher,* January 1973, **66:** 27–28.

46. John Glenn. "The Quest for the Lost Region," *Mathematics Teaching,* 1968, **43:** 23–25.

47. Solomon W. Golomb. *Polyominoes.* New York: Charles Scribner's Sons, 1965.

48. Vincent H. Haag, Clarence E. Hardgrove, and Shirley A. Hill. *Elementary Geometry.* Reading, Mass.: Addison-Wesley, 1970.

49. Arthur E. Hallerberg. "The Metric System: Past, Present — Future?" *The Arithmetic Teacher,* April 1973, **20:** 247–255. (Other selective readings on the metric system are included in this journal.)

50. D. Hilbert and S. Cohn-Vossen. *Geometry and the Imagination.* New York: Chelsea, 1952.

51. Alan Holden. *Shapes, Space and Symmetry.* New York: Columbia University Press, 1971.

52. Hilda P. Hudson. *Ruler and Compass,* London: 1916. New York: Chelsea, 1953. (Reissue; bound with A. B. Kempe: *How to Draw a Straight Line;* E. W. Hobson: *Squaring the Circle;* and so on.)

53. H. E. Huntley. *The Divine Proportion.* New York: Dover Publications, 1970.

54. Max Jeger. *Transformation Geometry*. New York: American Elsevier, 1969.

55. Donald E. Jennings. "An Intuitive Approach to Pierced Polygons," *The Mathematics Teacher,* April 1970, **63:** 311–312.

56. Donovan A. Johnson. "Paper Folding for the Mathematics Class." Washington, D. C.: National Council of Teachers of Mathematics, 1957.

57. Mervin L. Keedy and Charles W. Nelson. *Geometry — A Modern Introduction.* Reading, Mass.: Addison-Wesley, 1965.

58. Richert Kelley. Elementary Mathematics for Teachers. San Francisco, Calif: Holden Day, 1970.

59. Maurice Kingston. "Mosaics by Reflection," *The Mathematics Teacher,* April 1957, **50:** 280–286.

60. Mary Laycock. *Straw Polyhedra.* Palo Alto, Calif.: Creative Publications, 1970.

61. W. Lietzmann. *Visual Topolgy.* London: Chatto and Windus, 1969.

62. Harry Lindgren. *Geometric Dissections.* Princeton, N. J.: D. Van Nostrand Company, 1964.

63. Caroline H. MacGillavry. *Symmetry Aspects of M. C. Eschers Periodic Drawings.* Utrecht: A. Oosthoek's Uitgeversmaatschappig NV, 1965.

64. J. E. Mann. "Discovering Special Polyhedra," *Mathematics Teaching,* Winter 1969, **49:** 48–49.

65. R. McWeeny. *Symmetry: An Introduction to Group Theory and Its Applications.* New York: Pergamon Press, 1963.

66. Rochelle Wilson Meyer. "Mutession: A New Tiling Relationship Among Planar Polygons," *Mathematics Teaching,* Autumn 1971, **56:** 24–27.

67. Minnesota Mathematics and Science Teaching Project, *Introducing Symmetry,* University of Minnesota, 1968.

68. Edwin E. Moise. *Elementary Geometry from an Advanced Standpoint.* Reading, Mass.: Addison-Wesley, 1974(1963).

69. Josephine Mold. *Topics from Mathematics — Solid Models.* New York: Cambridge University Press, 1967.

70. Josephine Mold. *Topics from Mathematics — Tessellations.* New York: Cambridge University Press, 1969.

71. H. M. J. Monie. "Tessellating Hexagons," *Mathematics Teaching,* Spring 1971, **54:** 26–27.

72. Charles Moore. "Pierced Polygons," *The Mathematics Teacher,* June 1968, **61:** 31–35.

73. Walter Myer. "Garbage Collecting, Sunday Strolls, and Soldering Problems," *Mathematics Teaching,* April 1972, **65:** 307–309.

74. National Council of Teachers of Mathematics. *Geometry in the Mathematics Curriculum.* Thirty-sixth Yearbook, Reston, Virginia: National Council of Teachers of Mathematics, 1973.

75. National Council of Teachers of Mathematics. *Multisensory Aids in the Teaching of Mathematics.* Eighteenth Yearbook. Washington, D. C.: National Council of Teachers of Mathematics, 1945.

76. National Council of Teachers of Mathematics. *Twentieth Yearbook on National Council of Teachers of Mathematics—The Metric System of Weights and Measures.* Teachers College, Columbia University, New York: Bureau of Publications, 1948.

77. James R. Newman. *World of Mathematics.* New York: Simon and Schuster, 1956. Vol. 1, 2, 3, 4.

78. Nuffield Mathematics Project, *Environmental Geometry,* New York: John Wiley & Sons, 1968.

79. Nuffield Mathematics Project, *Shape and Size* (No. 2 & No. 3), New York: John Wiley & Sons, 1968.

80. Oystein Ore. *Graphs and Their Uses.* New York: Random House, The L. W. Singer Company, 1963.

81. J. P. Phillips. "The History of the Dodecahedron," *The Mathematics Teacher,* March, 1965, **58:** 248–250.

82. Joseph A. Raab. "The Golden Rectangle and Fibonacci Sequence, as Related to the Pascal Triangle," *The Mathematics Teacher,* November 1962, **55:** 538–543.

83. H. Rademacher and O. Toeplitz. *The Enjoyment of Mathematics.* Princeton, N. J.: Princeton University Press, 1957.

84. Ernest Ranucci. "Jungle-Gym Geometry," *The Mathematics Teacher,* January 1968, **61:** 25–28.

85. Ernest R. Ranucci. "Master of Tessellations: M. C. Escher, 1898–1972," *The Mathematics Teacher,* April 1974, **67,** No. 4: 229–306.

86. Ernest R. Ranucci. "Space Filling in Two Dimensions," *The Mathematics Teacher,* November 1971, **64:** 587–593.

87. Ernest R. Ranucci. "Tiny Treasury of Tessellations," *The Mathematics Teacher,* February 1968, **61:** 114–117.

88. Ronald C. Read. *Tangrams — 330 Puzzles.* New York: Dover Publications, 1965.

89. Garth E. Runion. *The Golden Section and Related Curiosa.* Glenview, Ill.: Scott Foresman, 1972.

90. William L. Schaaf. *A Bibliography of Recreational Mathematics.* Vols. 1 and 2, Reston, Virginia: National Council of Teachers of Mathematics, 1955 and 1970.

91. School Mathematics Project. *The School Mathematics Project.* Books A, B, C, D, E, and F, New York: Cambridge University Press, 1971.

92. Carol Elizabeth Shengle. "A Look at Regular and Semi-Regular Polyhedra," *The Mathematics Teacher,* December 1972, **65:** 713–718.

93. A. V. Shubnikov and N. V. Belov, et. al, edited by W. Holser. *Colored Symmetry.* Oxford: Pergamon Press, 1964.

94. David F. Siemens. "The Math of the Honeycomb," *The Mathematics Teacher,* April 1965, **58:** 334–337.

95. H. Steinhaus. *Mathematical Snapshots.* New York: G. E. Stechert, 1938.

96. Peter S. Stevens. *Patterns in Nature.* Boston, Mass.: Little, Brown, 1974.

97. Joseph L. Teeters. "How to Draw Tessellations of the Escher Type," *The Mathematics Teacher,* April 1974, Vol. 67, No. 4: 307–310.

98. D'arcy W. Thompson. *On Growth and Form.* London: Cambridge University Press, 1961.

99. Heinrich Tietze. *Famous Problems of Mathematics.* Baltimore, Md.: Graylock Press, 1965.

100. L. Fejes Toth. *Regular Figures.* Oxford: Pergamon Press, 1964.

101. Charles W. Trigg. "Collapsible Models of Regular Octahedrons," *The Mathematics Teacher,* October 1972, **65:** 530–533.

102. Marion Walter. "An Example of Informal Geometry: Mirror Cards," *The Arithmetic Teacher,* October 1966, **13,** No. 6: 448–452.

103. R. F. Wardcop. "A Look at Nets of Cubes," *The Arithmetic Teacher,* February 1970, **17:** 127–128.

104. Peter Wells. "Symmetries of Solids," *Mathematics Teaching,* Summer 1971, **55:** 48–52.

105. Magnus Wenninger. *Polyhedron Models.* New York: Cambridge University Press, 1970.

106. Magnus J. Wenninger. "Polyhedron Models for the Classroom," Washington, D. C.: National Council of Teachers of Mathematics, 1966.

107. H. Weyl. *Symmetry.* Princeton, N. J.: Princeton University Press, 1952.

108. John Winter. *String Sculpture.* Palo Alto, Calif.: Creative Publications, 1972,

109. R. C. Yates. *Geometric Tools: A Mathematical Sketch and Model Book.* St. Louis, Mo.: Educational Publishers, 1949.

The following are selected topics from Martin Gardner's "Mathematical Games" section of the *Scientific American.* The date and topic are given for each article.

June	'57	The Mobius Band
Dec.	'57	Polyominoes
Mar.	'58	Left- and right-handedness
Sept.	'58	The soma cube
Dec.	'58	Diversions which involve the Platonic solids
Jan.	'59	Mazes: how they can be traversed
July	'59	Origami
Aug.	'59	The golden ratio (Phi)
May	'60	The packing of spheres
June	'60	Paperfolding and papercutting
Sept.	'60	The four-colour problem
Nov.	'60	More about polyominoes
Apr.	'61	Coxeter's *Introduction to geometry*
Sept.	'61	Topological diversions
Nov.	'61	Dissections
May	'62	Symmetry and asymmetry
July	'62	Abbot's "Flatland" and two-dimensional geometry
Nov.	'62	Checker-board puzzles: dissections, etc.
Feb.	'63	Curves of constant width
May	'63	Reptiles
July	'63	Topological diversions
Oct.	'64	Simple proofs of Pythagoras

Dec. '64   Polyiamonds

Feb. '65   Tetrahedrons

July '65   "Op Art" patterns: tessellations

Oct. '65   Pentominoes and polyominoes

Dec. '65   Magic stars, graphs and polyhedrons

Apr. '66   The eerie mathematical art of Maurits C. Escher

Dec. '66   Pascal's triangle

June '67   Polyhexagons and polyabolos

July '67   "Sprouts" and "Brussels sprouts": topological games

Dec. '68   The Mobius strip

Mar. '69   The Fibonacci sequence

Sept. '69   Constructions with a compass and a straightedge

May '70   Optical illusions

June '70   Elegant triangle theorems not found in Euclid

Apr. '71   Geometric fallacies: hidden errors pave the road to absurd conclusions

Sept. '71   The plaiting of Plato's polyhedrons and asymmetrical Ying-Yan

Apr. '72   A topological problem and eight other puzzles

Sept. '72   Pleasurable problems with polycubes

Dec. '72   Knotty problems with a two-hole torus

Jan. '73   New games: Sim, Chomp, and Race Track

Sept. '73   Problems on the surface of a sphere

Feb. '74   Cram Crosscram and Quadrathage: new games having elusive winning strategies

June '74   Dr. Matrix, numerology, and pyramids

July '74   On the patterns and unusual properties of figurate numbers

Aug. '74   On the fanciful history and creative challenges of the puzzle game of tangrams

Sept. '74   More tangrams: combinatorial problems and the game possibilities of snug tangrams

July '75   On tessellating the plane with convex polygons

Aug. '75   More about tiling the plane; the possibilities of polyominoes, polyiamonds and polyhexes

PICTURE CREDITS

The authors would like to thank the following persons, institutions, and publishers for their cooperation in supplying the photographs reproduced in this book.

Chapter 1. Page 4: Nautilus shell, Honeycomb, Grant Heilman; Fern leaf, Lester V. Bergman and Associates. Page 5: Snowflake, Patty Benner; Pine cone, Grant Heilman; Sunflower, Robert Houser, Rapho Division, Photo Researchers, Inc. Page 6: Snail, Grant Heilman; Parthenon, Editorial Photocolor Archives. Page 7: Starfish, Patty Benner; Sunflower, Grant Heilman. Page 8: Arrowheads, Mimbres bowl, The Peabody Museum, Harvard University; Baskets, KC Publications. Page 9: Pyramids at Giza, Editorial Photocolor Archives. Page 16: *St. Jerome,* Leonardo DaVinci, Scala Fine Arts Publishers; *Broadway Boogie Woogie,* Piet Mondrian, 1942-43. Oil on canvas, $50'' \times 50''$. Collection, The Museum of Modern Art, New York. Page 17: *Verbum,* M. C. Escher, Escher Foundation, Haags Gemeentemuseum, The Hague; Navaho rug, Two Gray Hills Rug, Daisy Taugelchee, KC Publications; Steeplechase quilt, Elizabeth Ann Cline, 1865, Denver Art Collection, gift of Mrs. Charlotte Jane Whitehill; Modern sculpture, *CUBI III,* David Smith, Scala Fine Arts Publishers; African sculpture, Marc and Evelyne Bernheim, Woodfin Camp.

Chapter 2. Page 26: Spiral pattern, Los Alamos Scientific Laboratory, University of California; Patterns of lines connecting 24 points, *The Story of Mathematics,* Hy Ruchlis and Jack Engelhardt, Harvey House; *Altair Designs,* Ensor Holiday, copyright © 1970 by Altair. Reprinted by permission of Pantheon Books, a Division of Random House, Inc.

Chapter 3. Page 74: Greek and Egyptian patterns, *Regular Features,* L. Fejes-Toth, Pergamon Press, Inc. Page 75: Chinese patterns, *Regular Features,* L. Fejes-Toth, Pergamon Press, Inc. Page 76: Architectural patterns at Mitchell Gardens, Daniel Brody, Stock, Boston Inc. Page 77: Alhambra drawings, M. C. Escher, Escher Foundation, Haags Gemeentemuseum, The Hague. Page 78: *Day and Night,* M. C. Escher, Escher Foundation, Haags Gemeentemuseum, The Hague. Page 104: Two drawings by M. C. Escher from *Symmetry Aspects,* Drawing 1946 (horsemen) and Drawing (frogs), Escher Foundation, Haags Gemeentemuseum, The Hague.

Chapter 4. Page 110: Transamerica Corporation; Radiolarians, *On Growth and Form,* D'arcy Wentworth Thompson, Cambridge University Press. Page 111: Pentagon, United States Navy. Page 112: Three photographs of walls of bricks, *Mathematical Snapshots,* 3rd edition, H. Steinhaus. Copyright © 1950, 1960, 1968, 1969 by Oxford University Press, Inc. Reprinted by permission. Page 119: Fluorite crystal, Lester Bergman and Associates; cubic crystal, Walter Dawn. Page 120: Radiolarians, *On Growth and Form,* D'arcy Wentworth, Cambridge University Press. Page 122: Polyhedron with octahedron, *Mathematical Snapshots,* 3rd edition, H. Steinhaus. Copyright © 1950, 1960, 1968, 1969 by Oxford University Press, Inc. Reprinted by permission. Page 129: Madagascar periwinkle, Walter Dawn; Snowflake, Patty Benner. Page 137: Blocks and truncated blocks, *Shapes, Space and Symmetry,* Alan Holden, Columbia University Press, New York, 1971. Page 140: Semi-irregular solids, *Shapes, Space and Symmetry,* Alan Holden, Columbia University Press, New York, 1971. Page 150: Cube with square-based pyramids, tessellation of rhombic dodecahedron, *Mathematical Snapshots,* 3rd edition, H. Steinhaus. Copyright © 1950, 1960, 1968, 1969 by Oxford University Press. Reprinted by permission.

Chapter 5. Page 165: Drawing by Richter; Copr. 1957 The New Yorker Magazine, Inc. Page 175: Drawing by W. Miller; © 1962 The New Yorker Magazine, Inc. Page 176: Drawing by Chas. Addams; Copr. 1957 The New Yorker Magazine, Inc. Page 177: *Magic Mirror,* M. C. Escher, Escher Foundation, Haags Gemeentemuseum, The Hague.

Chapter 7. Page 234: Assembly line, courtesy of Chevrolet Motors Division, General Motors Corporation. Page 235: Model of L-1011 TriStar jetliner and full-size finished product, courtesy of Lockheed-California Company; Aerial views, courtesy of Itek Corporation.

Chapter 9. Page 287: X-ray diffraction patterns, courtesy of Dr. B. E. Warren; virus crystal, courtesy of The Public Health Research Institute of the City of New York, Inc.; constellation of Orion, Hale Observatories. Page 299: top left and right, from *String Sculpture,* John Winter, published by Creative Publications, Inc., Palo Alto, California 94303; bottom right, *Mathematical Snapshots,* 3rd edition, H. Steinhaus. Copyright © 1950, 1960, 1968, 1969 by Oxford University Press, Inc. Reprinted by permission.

Chapter 10. Page 316: Egyptian mural, Lehnert & Landrock, Cairo.

Index